Rewilded

Saving the South China Tiger

野化拯救华南虎

Rewilded

Saving the South China Tiger

野化拯救华南虎

（汉英对照）

By Li Quan 全 莉 著

北京出版集团公司

北京出版社

Save China's Tigers

First published in the United Kingdom in 2010
By Evans Mitchell Books
54 Baker Street, London W1U 7BU, United Kingdom
info@embooks.co.uk
www.embooks.co.uk

Author: Li Quan
Design: Clare Mellor at fullstopdeisgn.co.uk
Tiger illustrations: Jordan Kirtchev
Origination, printing and binding: CandCprinting.com, China

British Library Cataloguing in Publication Data. A CIP record of this book is
available on request from the British Library.

Productions costs kindly donated by Conservation Finance International.
All Profits from sales of this book benefit 'Save China's Tigers' charity.

ISBN 978-1-901268-54-6

Contents

虎，是百兽之王。

古老的中国有许多传说中的祥禽瑞兽，比如龙，凤，麒麟，鲲鹏，狻猊，貔貅……但堪称图腾且真实存在的动物，唯有虎。高悬的年画，娃娃的帽头，调兵遣将的虎符，虎贲骁将的旌旗，还有无数中国人的姓氏与名讳，民谚和成语，山川加地名……虎，在我们的生活中随处可见，贯穿了中华文化的精髓。

虎，从远古时代的"中国古猫"进化而来。当今世界上九个亚种的虎，都是我们的"中国老虎"——华南虎的后裔。但就在这短短的几十年时间里，华南虎已经从山林野外消失，仅存的几十只也都关在动物园里，种群衰减，个体退化，繁殖萎缩。这种让中国人喜爱和骄傲的百兽之王，超过两百万年进化而来的美丽生灵，就要在一二十年内灰飞烟灭！谁，能在此时伸出手来，挽救悬崖边上的中国老虎？

全莉，我大学时代的小师妹，成了华南虎的救星。她北大毕业后又深造于美国沃顿商学院和劳德学院，之后在时尚界闯出一片天地，先后做到贝纳通和古琦等国际大品牌的高管。但这个生于虎年的女孩没有忘记我们的中国虎。当某些权威国际野生动物组织宣布华南虎的仅存数量难以维持繁衍、该物种注定灭绝时，全莉拍案而起。她离开了高贵奢华的时尚界，毅然走向苍凉粗砺的南非草原。

全莉的想法是：既然华南虎在动物园的铁笼中难逃灭绝厄运，何不放虎归山，将圈养的老虎暂时放到能够回归野性的地方呢？目前的中国虽然还不具备这个条件，但南非无疑是一个绝佳的选择。全莉在她的银行家丈夫斯图尔特·博锐的帮助下注册了"拯救中国虎国际基金会"，在南非购买下17个彼此毗连、共占地330平方公里的倒闭农场，引进了多种土生土长的猎物，成为野化华南虎的理想平台。全莉希望，一旦中国也建成保护区，野化成功的中国虎就能从这里源源不断地回归祖国，继而中国也可以开始本土的野化，从而拯救华南虎这一物种……

但接下来她面临的竟是无休止的批评，甚至遭到有地位的野保人士的恶意敌对，他们固守失败的野保方式，对全莉百般讽刺。他们说：全莉是在发疯！把老虎从动物园转送到另一个大陆的陌生环境，这是多大的风险？老虎活不了。他们认定全莉的做法非但救不了华南虎，反而会加速它们的灭绝。

但全莉认准的事，她会坚持到底。同时她的计划也得到了那些锐意进取、不墨守陈规的生态专家和野保人士尤其是中国政府的支持。2003年9月2日是个值得纪念的日子，这一天，首批来自中国的华南虎幼崽"希望"和"国泰"飞抵南非，走下机舱，迈出铁笼，试探着踏上了这块从未见过的非洲大地。

它们能生存下来吗？它们能适应这里的水土，学会打猎，养活自己，生儿育女，繁延血脉，最终完成拯救这个物种的伟大使命吗？让我们翻开这本书，跟着全莉和她的团队，重新体验一遍这八年多年来的辛酸喜乐吧……

INTRODUCTION
By Justin Wintle

It was mid-October, 2003. As the sun descended over the South African veldt, Li Quan braced herself. She was about to discover whether everything she had fought so hard for so long was really coming to fruition, or whether all her plans were as nought. As the long transport vehicle entered the camp created on the scrubland she involuntarily held her breath. The truck was bringing two covered cages. And she continued holding her breath until slowly, very slowly, the door of the first cage opened. For a minute or so, an agonizing eternity, nothing happened. Then a head appeared, gingerly sniffing the warm air, and surveying its surroundings with the utmost wariness.

Hope had arrived, Hope being a South China tiger cub. Brought by air and road from a zoo thousands of miles away in China itself, the cub suddenly sprang out of the cage and took his first, tentative steps on the African soil at his feet. Until this moment all he had known were the concrete floors of zoos and quarantine cells, and the bars that kept him in.

Soon enough a second cub, Cathay, followed Hope out into the open. Thanks to Li Quan and her team, the little big cats were about to embark on the discovery of many freedoms hitherto forbidden them -- the freedoms to roam, play, eat, sleep, hunt and yes, even to kill in a completely natural environment where humankind intervened only for their betterment.

In short, Hope and Cathay were to be 'rewilded', and have their natural animal dignity restored to them.

But it was not just an act of kindness toward two individual South China tiger cubs that had compelled Li Quan to bring them to a place where land -- real wild land -- awaited them, and where they could be protected as well as set at liberty. Li entertained a much greater ambition. By creating a South China tiger colony in South Africa she aimed to save the very species itself. For the sad, tragic truth was that in their native land the South China tiger population had shrunk to less than a hundred -- all or nearly all of them behind bars in zoos, which did not encourage them to breed.

For Li Quan the South China tiger was worth any amount of effort, time and resources to spare. Not only was it a magnificent predator in its own right, and the commonly acknowledged ancestor of all other species of tiger, but as the King of the Hundred Beasts it held a long and honoured place in her own Chinese culture and history.

The earliest fossilised remains of the South China tiger date back two million years. For Li, it was unbearable to think that two million years could be wiped out in a mere decade or two.

Born in the Year of the Tiger -- when else? -- and raised in China's capital Beijing, Li had travelled to the United States to further her education at the Wharton Business School and the Lauder Institute. Soon she was embarked on a successful career in the fashion world, working as an executive for both Benetton and Gucci. But she had not forgotten the iconic South China Tiger. When some of the world's most influential and powerful wildlife organizations started saying South China tigers were doomed, that there were too few of them to rebuild a viable species, she became enraged. Who were these desktop conservationists anyway? And had anybody actually tried to save this most particular subspecies of tiger?

So she quit her career and put her whole energy, her whole being, into finding a practical means of pulling the South China tiger back from the brink of the bottomless, black chasm of extinction.

She had had a brainwave. If the South China tiger was extinct, or almost extinct, in the rapidly dwindling plains and forests of China itself, why not take some of those that were in captivity somewhere they could rediscover what being wild means? And if that could be done, then surely they would begin breeding properly again?

South Africa was the obvious choice. Eventually, Li Quan hoped, enough South China tigers could be returned to China, once protected reserves had been created for them. For now, though, she faced no end of criticism and sometimes outright hostility from established conservationists, stuck with their own shibboleths, their own way of doing things. She was mad, they said, no tiger would survive the transition from zoo to a wholly alien environment on another continent where who knew what dangers awaited them. Most probably, they hinted, Li Quan would hasten the end of the South China tiger, rather than prolong its existence.

But Li Quan was nothing if not persistent. She enlisted the support and advice of some key conservationists prepared to think outside the box of conventional conservation strategies, and, critically, she persuaded both the Chinese and South African governments that her plan could, indeed would work.

And she had one other trump card up her sleeve, her husband Stuart Bray.

They had met as MBA students in America. Since then Bray had pursued his own glittering career, becoming a much respected investment banker. But like his wife, he too had become disillusioned with the politics of office life, and so he pledged his full commitment to her tiger project. With his help, Li Quan was able to establish Save China's Tigers as an international charitable organisation registered in Britain, the US, Hong Kong and Australia. And together they were able to acquire seventeen defunct but adjacent sheep farms in South Africa, where respect for big cats and other larger wildlife was ingrained in the national mindset.

Between them, the old farms made up some 300 square kilometres of overgrazed land. Once made secure and stocked with plentiful local prey, this became the perfect platform for rewilding the South China tiger.

It would have taken much longer, probably too long, to create a similar nature reserve in China, where conservation programs were still in their infancy. That would have to come later. For Li, the endgame remains when a sizeable colony of South China tigers can return home, to a safe but nonetheless wild habitat.

A fast-track solution was the only possible solution. Make it happen, Li had said to herself over and over year in year out, make it happen! And now, suddenly, at the end of a warm October afternoon far far away from where either Li or the cubs had been reared, it seemed it actually, really was happening.

As Hope and Cathay began exploring their new habitat, carefully prepared for them with the help of seasoned ecologists and biologists, with a confidence that surged by the minute, Li's heartbeat quickened, Yes! It could work, it was already working! Hope and Cathay were the living proof before her very eyes. Survival was embedded in their genes.

Her delicate Chinese fists clenched with joy. Her critics were wrong. They were dull, unimaginative people who dared not dream. But she did not allow her satisfaction to give way to stronger feelings of triumph. She was too canny for that. She knew that the cubs' arrival at the rewilding center was a beginning, not an end. October 9th 2003 was a momentous day for sure, but even harder work lay ahead, if what by any reckoning was a uniquely audacious conservation venture were to pan out.

Seven years on, the omens look good, despite some inevitable set-backs. More tigers have been introduced from China, and five Laohu Valley-born young tigers are hunting efficiently on their own. The South China tiger colony is growing, putting first one stealthy paw forward, then another, then another.

This book is a photo-documentary of what has already been achieved to preserve the South China tiger, and a testimony to the courage and determination of Li Quan and her many helpers.

A longstanding friend of Li Quan and Stuart Bray, Justin Wintle is a British author, journalist and broadcaster whose many books include Romancing Vietnam, Furious Interiors: Wales, R.S.Thomas and God and Perfect Hostage, his challenging biography of the Burmese Nobel Peace Prize winner Aung San Suu Kyi.

1. Arriving in Africa
第一章 抵达非洲

First South China Tigers in Africa: Cathay & Hope
华南虎初踏非洲

Diary: TigerWoods and Madonna's Adventurous Journey
日记: 虎伍兹和麦当娜南非冒险之征

Here Comes '327'
327来了

From the cages of Shanghai Zoo to the grasslands of South Africa is a long, long way. Cathay and Hope, 9 months and 8 months respectively, were the first two cubs to make that journey, traveling 'first class' courtesy of Cathay Pacific Airways. After nearly two days they arrived safely, on 2 September 2003. To comply with the law they had necessarily to spend the next five weeks in quarantine. During this time they were treated for ringworm, as well as receiving dental care. Then in October, after a lifetime walking on concrete, Hope and Cathay finally stepped onto the natural earth of the grasslands. It took them a little time to get used to a natural environment, but soon enough they settled down into their new life.

A year later, almost to the day, two more cubs, Madonna and TigerWoods, made the same long journey to the Laohu Valley Reserve, so-called because Laohu means 'tiger' in Chinese. The voyage and the sun took their toll on Madonna, and for a while it seemed she might not survive. But duly she recovered, and joined her mate TigerWoods for their rewilding training.

In April 2007 a four year old male tiger arrived from Suzhou Zoo. His name was '327', his studbook number in the registry of South China Tiger kept by the Chinese Zoological Association. Intended as a hot stud to the maturing tigresses, city cat 327 initially had reservations towards his new surroundings. Soon, though, he was devouring game with a ferocious appetite.

从上海动物园的铁笼到南非的草原，两只分别为9个月和8个月大的"国泰"和"希望"是首批乘坐国泰航空公司头等舱不远万里、长途跋涉来南非的虎崽。经过两天的征途，他们终于于2003年9月2日平安抵达。根据相关法律与规定，两只幼崽接受了长达一个月的检疫，这期间还接受了对金钱癣和损伤牙齿的治疗。一生下来就在水泥地上生活的幼崽们，终于踏上了草原的天然土壤。经过短时间的过渡适应自然环境后，两只幼崽开始了崭新的生活。

一年后，即2004年10月，另外两只幼崽"麦当娜"和"虎伍兹"也踏上了漫长的旅程，来到南非一个以中国老虎命名的保护区。万里长征及炙热的太阳造成麦当娜身体不适，一时令人担心她不能存活下来。但她恢复了健康，与小伙伴虎伍兹开始了野化训练的历程。

2007年4月，从苏州动物园来了一头4岁的公虎，他叫"327"，这是中国动物园协会为他注册的华南虎血统证书编号。327是为逐渐成熟的母虎们做配偶而来。可城市中长大的他最初对新环境却不以为然，但对野味兴趣极大，胃口极好。

Top left: Cathay
左上：国泰
Top right: Madonna
右上：麦当娜
Bottom left: TigerWoods
左下：虎伍兹
Bottom right: Hope
右下：希望
Right: '327'
右：327

First South China Tigers in Africa: Cathay & Hope
华南虎初踏非洲

Cathay, Born on 21 January 2003 in Shanghai Zoo

国泰：2003年1月21日出生于上海动物园

Hope, Born on 17 February 2003 in Shanghai Zoo

希望：2003年2月17日出生于上海动物园

1. Cathay and Hope were both born in Shanghai Zoo, in January and February 2003 respectively. After spending a month in quarantine, on 1 September they left for Beijing International Airport and then flew to South Africa via Hong Kong.

国泰和希望2003年1月和2月先后在上海动物园出生。经过一个月隔离检疫后，9月1日从北京国际机场经香港飞往南非。

2. After thirteen hours, they arrived safely in Johannesburg.

13个小时提心吊胆的旅程后，他们平安抵达南非约翰内斯堡。

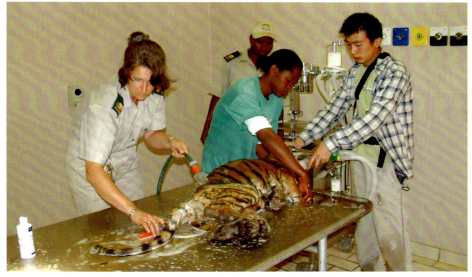

3. They spent over a month in quarantine again.

During quarantine red spots appeared on Hope's body and his coat became patchy. He had developed ringworms and was treated immediately. Ringworm is common in captive conditions.

两只幼虎又在南非接受了一个多月的隔离健康检疫。

检疫期间，发现希望身体出现非常明显的红斑,背毛零乱。经诊断为金钱癣，并用药物溶液为他们治疗。在动物聚集的场所金钱癣是一种常见病。

4. Hope's broken canine and Cathay's broken tooth and canine, arising from biting metal bars during the travels, were extracted.

希望断了的犬齿、国泰的断齿与3颗破裂的犬齿也被取了出来，这都是旅行中咬铁笼造成的。

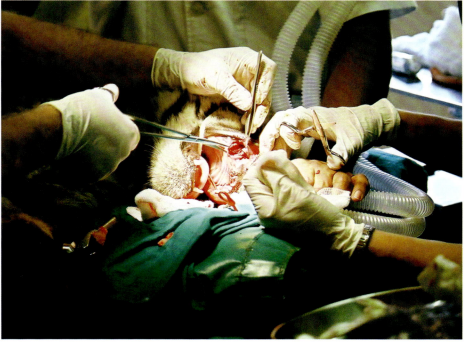

5. After quarantine they started their rewilding training.

检疫结束后，他们开始了野化训练。

6. Hope's transport cage was opened but he hesitated for several minutes before dashing out. Seeing that, Cathay then followed.

希望的运输箱先被打开，但他战战兢兢不肯出来，犹豫了足有几分钟才猛然窜出笼子。国泰见状，也壮胆跑出箱子。

7. Zoo-born tigers are like city kids. The cubs refused to leave the small concrete area to step onto African soil. But Cathay had come with a toy basketball. When she saw her basketball flying in through the air she chased after it, forgetting the scary earth. Problem solved!

可是动物园里生长的老虎，如同城市里的儿童，小虎们不肯从水泥地下到土地上去。人们只好想办法。国泰有个玩具篮球，当他看到心爱的篮球飞进来时，噌的一下就冲了上去，忘了对土地的恐惧。

8. The cubs looked upon a feathered chicken as a toy. Later they would turn their noses up at sliced chicken meat.

小虎们把带毛的鸡当成玩具，对拔了毛刨开的鸡没有兴趣，切好的两片肉连闻都不闻。

9. They had been on a long long journey. The little ones needed some well-deserved rest before rewilding training started.

路上也的确辛苦了，让他们在新家先缓缓神儿，休息休息吧。

10. Several days later Cathay chuffed at Li Quan for the first time, as if to thank her for bringing them out under the open sky. Chuffing is a sound a tiger makes through its nose to express friendliness and greetings.

几天后，国泰对全莉发出了虎特有的"扑哧"声，像是感激她把他们带进了大自然。

Diary by Li Quan: TigerWoods and Madonna's Adventurous Journey

全莉日记：虎伍兹和麦当娜南非冒险之征

TigerWoods, Born on 9 March 2004 in Shanghai Zoo
虎伍兹：2004年3月9日出生于上海动物园

Madonna, Born on 20 April 2004 in Shanghai Zoo
麦当娜：2004年4月20日出生于上海动物园

29 October to 6 November 2004

It did not take long to get TigerWoods and Madonna into the purpose-made transportation crates at Beijing Zoo and loaded into a truck ready for the airport. However, being confined prompted them to growling, roaring and throwing themselves against the sides of the crates.

The Cathay Pacific flight was scheduled for 07.50 on 29 October. Representatives of the cargo handling agent gave the cubs priority transit. The cubs, especially TigerWoods, were not very appreciative. He roared and was unsettled. TigerWoods and Madonna were mother-reared at Shanghai Zoo and had very little contact with humans. In fact when they first arrived in Beijing Zoo for quarantine on 27 September, any sight of humans would cause them to hiss, roar and jump up and down. They calmed down somewhat during the month of quarantine.

The flight arrived in Hong Kong around noon. The connecting flight to Johannesburg was at 23.45 that evening. The cubs were taken into the cargo area for animals.

We barely managed to put some water into the two small water tanks in their crates before the flight was due to depart, only to discover that TigerWoods had bitten a hole in one of the iron tanks and it was now leaking.

Nothing could stop me worrying during the long 14 hour flight.

As we landed in Johannesburg I phoned our South African team on standby and waiting for the State Vet to check all the required documentation for the cubs importation. I was greatly relieved when they said they had now seen the cubs.

It was a long wait for the cubs to be cleared through customs. The Vet was, first of all, late. When he finally arrived he said that the Chinese

2004年10月29日到2004年11月6日

在北京动物园，没费多大工夫就将两只小虎"虎伍兹"和"麦当娜"装进了特制运输木箱中，抬上卡车，准备运往机场。小虎们更加不满，不时低吼、咆哮，撞击箱子壁。

国泰航空317次航班定于10月29日早上7：50起飞。在机场货运代理处，代理处的负责人给了小虎们特殊待遇，优先检查过关。但是两只小家伙，特别是"虎伍兹"并不很感激，他不停地吼叫，在箱里焦躁不安地移动。"虎伍兹"和"麦当娜"在上海动物园一直都是由母虎带着的，之前很少接触人类。当他们9月27日刚到达北京动物园的时候，只要一见人，就会焦躁不安，咝咝地叫着，咆哮，上下乱撞乱跳。在一个月的检疫期间，才慢慢安静点儿了。

我们中午12点到达香港。到约翰内斯堡的转接航班在晚上11点45起飞。小虎们被放进了专门存放动物的货运处。

飞机快起飞前，我们才成功地为小虎笼子里的容器放了一些水，但立即发现，"虎伍兹"把铁皮水盒咬了个洞，正在渗水。

我很担心在漫长的14小时里会发生什么。

10月30日早上7点到达约翰内斯堡国际机场时，我马上给前来接机的南非团队打电话问询。因为他们要等待国家兽医检查小虎入境所必须的文件。听到他们说已经见到的"异常活泼"的小虎们，我松了口气。

小虎出关花费了很长的时间。首先，国家兽医迟到了，好不容易等到他来，他却告诉我们，中国的兽检证书上没有明确地写出小虎接受过犬恶丝虫的检疫，尽管这次文件与去年一模一样，但那次我们运"国泰"和"希望"时没有遇到任何麻烦。而这位兽医却拒绝在小虎入境许可上签字，除非我们为小虎再注射一次犬恶丝虫疫苗。问题是这位兽医自己不肯为小虎注射疫苗，又逢星期六，找兽医哪那么容易。可我决不能容忍小虎再在木

veterinary certificates did not specify that the cubs were treated with heartworms even though this document was exactly the same as last year's and we had no problem getting Cathay and Hope through customs then. However, this vet refused to sign the approval paper to let the cubs into the country, unless we gave the cubs another heartworm injection. But he himself would not do the injection, and on a Saturday it would not be easy to find a vet. I couldn't let the cubs stay another day in the crates!

Petri Viljoen, Executive Director of our SA operation, deployed his magic immediately. After about an hour he managed to find the Deputy Director of the State Vet Department, Richard. We and the cubs took off in my friend Eddie Keizan's private Caravan at 12.15 for Laohu Valley Reserve in Philippolis Free State.

The cubs were taken to the 200 m². quarantine camp where Chinese Ambassador Liu Guijin had been waiting. As soon as the crate door was opened Madonna dashed out running along one side of the camp like mad. She must have felt disoriented having never been outside a cage larger than a few square metres. She stood up at one corner of the fence and having failed in her attempt to climb she ran along another side to the far corner.

TigerWoods on the other hand refused to come out of the cage, not until we had all left. A couple of hours later, TigerWoods was seen lying with his belly up and rolling on the ground, enjoying the African sun.

What a great relief it was to see both cubs now in their new spacious home!

The cubs did not seem to know the damage the strong African sun

箱子里多待一天！

中国虎南非项目中心执行总裁维纶此时大显身手，一小时之后，他设法找到了国家兽医局副主任查德，我们与小虎终于在12点15分乘朋友凯赞的私人飞机前往位于自由省菲利普里斯镇的老虎谷保护区。中国驻南非大使刘贵金早已在老虎谷保护区的检疫营地等待小虎的到来。箱门打开时，"麦当娜"从箱子里奋勇地冲了出来，在200平方米的营地中奔跑着。她肯定糊涂了，因为她从小至今都一直生活在几米大的笼子中，她在围网的一角踮着脚站起来，往上蹦，往上爬，失败后，又跑到了围栏的另一个角落里。

而在另一边的"虎伍兹"则拒绝从笼中出来，非得我们全部撤离。

两个小时之后，才看到"虎伍兹"肚子朝天地躺在地上打着滚儿，享受着南非的阳光。

能看到两只小虎来到宽广的新家真是莫大的安慰啊！

接下来的三天，小虎们大部分时间睡觉倒时差。看来他们不知道南非强烈的阳光的危害，拒绝到遮棚下面去。

我们高兴地发现留下的牛肉被吃掉了。

11月3日，麦当娜呕吐出大量的液体，虚弱的身体颤抖着，几乎不能站立。她的症状是脱水，接连几天的艳阳可能是罪魁。

我把一碗水推到麦当娜的面前，她却连看它一眼的力气都没有。我往她头上洒了些水，她这才慢慢地转过头来看到了水碗。她试图把头挪近水碗，但她太虚弱了，即使水碗离她才这么点距离她也没有力气够到。于是我把水碗放到她的嘴底下，使她只要放低下巴就能喝到水。她一点一点地吸吮着，看上去衰弱无比。

兽医第二天才能来。我的团队轮流昼夜值班。

营地里有水槽，但来自狭窄动物园环境的小虎崽们可能相当不适应这200米大的营地。口渴的时候，也不知道走到水槽那里。

第二天早上（11月4日），她的状况似乎好些了，但还有一点颤抖。只要我给她水她就会喝。她似乎知道我们想帮助她恢复健康。我举起她的脚

could do and did not want to move under the shelter. The beef left for them was eaten, to our delight.

On 3 November Madonna started vomiting. Her body was shaking and she could barely lift herself up. She had all the symptoms of dehydration.

I pushed a bowl of water next to her face but she did not have the energy to notice it. I pushed the water bowl right under her mouth so she could just lower her jaw to reach the water. She then drank and drank, in tiny sips.

The vet could only come the next day. My team spent the night guarding her in turn. The next morning she looked a lot better but still shaky. Whenever I pushed the water bowl under her nose she would drink. She seemed to know however that we were responsible for making her feel better today so when I lifted up her paw to examine her wounds we had spotted in between her fingernails she did not complain at all. She also gave me a typical friendly tiger chuff, for the first time ever. As if to thank me her back touched my legs when she walked by. When I couldn't resist touching her back, not only did she not mind, but she also chuffed back at me. What a change in her attitude!

The vet team came in the afternoon. We didn't want to use anesthesia or sedatives on a weak Madonna, so we caught her with a blanket. We held her legs and arms. Her jaws were tied together with a wide cloth ribbon to prevent her from biting us. Her eyes were covered with a blanket so she could not see the movements surrounding her. She did not seem to struggle at all. The vet Gavin dripped a litre of liquid into her stomach and treated her damaged teeth and infected fingernails.

掌检查她的伤口、给她的爪子作标记的时候，她不仅没抱怨，还对我露出高兴的鼻息声。这是老虎特有的一种鼻音，用以相互表示友好和问候。好像是要感激我,她走动时用身体蹭我的腿。我激动地抚摸她背时，她不仅没有怨言，还又对我打出了友好的响鼻。她的态度变化多大呀。

下午兽医们赶来了。麦当娜太虚弱了，不能使用麻醉剂，所以他们用毯子扑住了麦当娜。一个人按后腿，还有两人各按她的一条前肢。麦当娜的嘴则被一条宽布条扎了起来以防她咬人。她的眼睛也被毯子覆盖住，防止她见到周围的一切而狂躁不安。她并未多加挣扎。兽医加文给她的胃灌了一公升的药，作了全面检查，发现她掉了一枚犬齿和一些小牙齿，这可能是由于紧张而啃咬铁笼所导致的。她的爪子也有些感染。兽医给她注射了抗生素，对牙齿和爪子进行了治疗，还打了一针以改善她的食欲。接下来的5天，我们得把药放在她的食物中让她服用。

治疗结束后，兽医们把布条从麦当娜的嘴上取下，然后松手退到了她的一边。麦当娜可能挺喜欢受到重视，特别是治疗时被我们轻轻地搔脸颊，所以现在竟然不动身。我们把毯子抽走时，她露出惊讶，咬住了毯子不放，然后不慌不忙地转过脸来，看到身后几个庞大的人类，这才决定起身，慢慢地走开了，好像什么事都没有发生过一样。

治疗结束以后我们终于放心了，希望一切都会好起来。

11月5日早晨，麦当娜只吃了一杯生鸡蛋和肉末混合物，虎伍兹则吃了两份羊肉和牛肉，一共4公斤。兽医让我们把药放在麦当娜的食物中，我们绞尽脑汁，但她啥都不吃。无计可施之下，保护区主任皮特决定射杀一只野兔给虎伍兹，免得他再抢给麦当娜特制的胡萝卜鸡汤。当然让小虎早日适应野生猎物也是相当重要的，这样他们就能在不久的将来将捉到的猎物与他们现在的食物联系起来。

皮特的确是个好射手。我们见到的第一只野兔就被他捉到了。皮特对我说："等着瞧吧，这只兔子会让虎伍兹大不一样的。"

我们把野兔扔给了虎伍兹，他一下抱住野兔，就好像这是他自己捕获

After Madonna's treatment was over we took the cloth ribbon off her mouth, relaxed our grasp and backed off to one side of her. However, Madonna may have enjoyed the attention that she simply would not move. When we pulled the blanket off her eyes she looked surprised and held the blanket with her teeth. When she turned her head around and saw us humans standing behind her, she decided to stand up and then walked away without any hurry, and as if nothing had happened.

By 5 November morning Madonna had only eaten a cup of raw egg and meat mixture while TigerWoods had been eating both his and her share of lamb and beef. We scratched our brains to find things she could eat as the medicine can only be taken with food. We failed miserably. Finally our Reserve Manager Peter decided to shoot a wild rabbit for TigerWoods to prevent him from eating Madonna's food which he was doing only so happily.

Peter said to me when he got the first wild rabbit, 'Just watch, TigerWoods would be a different tiger now when he gets his wild rabbit.'

We threw the rabbit at TigerWoods and he grabbed it as if he had caught the rabbit himself and immediately began eating it as if this was not his first ever game. Madonna showed no interests in her chicken soup but got up instead to walk over to TigerWoods, chuffing at him. TigerWoods growled at Madonna. Somehow the wild rabbit triggered

的一样，立刻吃起来，这只他生平第一次吃到的野生动物，对他来说竟是如此习惯。而麦当娜对她的鸡汤一点也没兴趣。相反，她起身边打招呼边走向虎伍兹。我们不能确定麦当娜是对虎伍兹还是对野兔感兴趣。出乎意料,虎伍兹对着麦当娜咆哮起来。自从他们10月30日到了老虎谷，他们一直相依为命，睡觉都蜷在一起。看来野兔的确激起了虎伍兹的野性本能。被虎伍兹嘞道了一通，麦当娜只好灰溜溜地回到了原来的位置卧下。

　　不论麦当娜究竟想要什么，我们都为她感到很遗憾。皮特决定再打一只野兔给麦当娜，看看她的反应。

　　今晚比早上要幸运得多。我们带着虎伍兹的跳兔和麦当娜的野兔回到了营地，麦当娜已经把汤碗打翻了。我们把野兔推到麦当娜的鼻下，她咬了咬就放弃了。皮特后悔没事先把它切开。这可是麦当娜的第一只野兔啊，而现在，再要把它拿走就难了。我们用棍子想把野兔掏出来，但麦当娜搂着野兔不放。好不容易才把野兔从她那儿取了回来，切开皮毛，暴露出嫩肉部分，又还给她。虽然麦当娜马上夺走了野兔，但她并没吃它的意思。

　　这个时候，虎伍兹已经把他的野兔吃光了。为了防止他再从麦当娜那里偷取食物，我们把跳兔扔给了他。同时决定明天要么再请兽医，要么向兽医购买些抗生素注射剂，麦当娜不吃东西，我们无法给她喂药。

　　早晨，第二只野兔不见了，我们深信是虎伍兹吃的。但跳兔没有动过。麦当娜今天早上又腹泻了，看上去状况不好，直发抖。

　　皮特带着药和麻醉枪来了。他把抗生素装进管筒，第二次成功地将带有抗生素的针剂射进了麦当娜的大腿。她有点儿震惊，挪到虎伍兹身边，

TigerWoods' wild instinct. Madonna returned to her old position.

　　We felt sorry for Madonna. Peter decided to hunt another rabbit to give to Madonna.

　　For some reason we were a lot luckier tonight than this morning. We arrived back with a springhare for TigerWoods and a rabbit for Madonna. We threw the wild rabbit at Madonna and she attempted to bite the rabbit but gave up very quickly. We weren't sure if this was hurting her teeth and Peter regretted that he had not cut it open for her. We used a stick to get it away from her but it took quite a bit of struggle as she hung onto it.

　　Meanwhile TigerWoods had finished his wild rabbit completely and to prevent him from stealing from Madonna we threw him the springhare. We also decided to either get the vet over again or fetch some injectable antibiotics from him as we failed to administer any medicine.

　　Next morning the second rabbit had disappeared and we believed TigerWood had eaten it. The springhare was untouched. Madonna again had diarrhea and appeared weak.

Enjoy Laohu Valley
老虎谷的乐趣

1. As they played a bird caught Cathay and Hope's attention. They jumped up to catch it. As if making fun of them the bird darted left and right. Suddenly, it fluttered right past Hope's head. Hope sprang round, but still couldn't catch it. As the bird flew away his face flooded with disappointment.

国泰和希望玩儿得正欢，看到小鸟，跳着追起来。小鸟有意捉弄他们，来回跳动。突然，小鸟从希望的头上掠了过去。希望四肢同时跃起，右前掌在空中出击，想抓住鸟。可惜小鸟飞走了，希望失望地望着天空。

2. Cathay knew it was impossible to catch the bird with a frontal attack, so she outflanked it with the stealth of a thief. But before she could reach it the bird flew off. It was her turn to experience the misery of failure.

国泰知道从正面抓不到鸟儿，便从侧面迂回包抄，一副贼头贼脑的样子。可她还未到跟前，鸟就飞走了，她只好抬头望着远飞的"玩具"。

3. There were bees making honey in a tree. Hope moseyed around watching, scheming perhaps to steal a little honey for himself. But as soon as the bees became aware of him they swarmed around him, frightening Hope into beating a hasty retreat.

一群蜜蜂正在一棵树里酿蜜，希望居然跑到树下，不停地转来转去，看上去是想掏蜂蜜。可他刚靠近，蜜蜂就一窝蜂地在他头上晃来晃去，吓得他不停地后退，咬牙切齿地示威。

4. Cathay was stalking a group of birds using grass as cover. The birds were playing at the water's edge. But however nimbly she made her approach she couldn't catch them. As soon as they saw her they spread their wings and rose out of reach.

国泰借着草丛的掩护，想袭击一群小鸟。鸟特别多，正在水里自由地戏耍着。国泰迈着谨慎的脚步，全神贯注地逼近它们。小鸟们还是发现了她，突然腾飞起来冲向天空。

5. The cries of vervet monkeys perched high in a tree attracted a tiger, who circled the tree looking up at them with curiosity. But what was he to do?

顺着一群黑长尾猴发出嘈杂的报警声，看到一只老虎正围着猴群所在的树打转，并好奇地向上看。

6. As she matured Cathay became less well-behaved and stopped waiting her turn during feeding. This infuriated Hope. Whenever she tried to get her teeth into the carcass he was eating he would snarl fiercely. One time he bared his teeth most menacingly at her and lay atop the blesbok to stop her getting at it. Undeterred, Cathay managed to get her teeth into the blesbok's rump and started feeding. What could Hope do? If he moved the chances were Cathay would drag the carcase away from under him.

日渐成熟，国泰不再安分守己，有时也在食物上分一杯羹。每到此时，希望就会一边恼怒地抓着国泰，一边呲牙咧嘴地压在食物上进行捍卫。有一次他真的恼了，露出牙齿看似凶猛。但国泰没被吓到，偷偷从下面美美地享用食物。这招使希望手足无措，因为他无法两头兼顾。如果起身，国泰更容易把食物拖走。

7. When the cubs discovered a leopard tortoise they turned it upside down and played with it like a football.

两只小虎发现了豹纹龟，把它掀倒玩了起来，还当球一样踢着玩儿。

8. TigerWoods loved stalking all sorts of prey, be it bird or mouse or motor vehicle! He loved playing with Madonna too, chasing and fighting. Rewilding was going to plan.

虎伍兹总是爱潜伏在各种"猎物"旁边，可能是鸟儿、老鼠，甚至是汽车。他还喜欢与麦当娜玩耍、追逐、打闹，这些都是学打猎的必修课。

9. TigerWoods and Madonna were hunting mice one day, staring intently into the grass, then sometimes pouncing. If their attention wandered, they showed an equal fascination with sticks. TigerWoods grabbed a metre long branch. He lay on his back juggling it in his paws, or playing tug-of-war with Madonna.

有一天，麦当娜和虎伍兹发现草丛中有老鼠，就一直盯着，然后突然扑上去。能吸引他们注意力的不是老鼠就是棍子。虎伍兹抓住一根大概1米长的棍子朝天仰卧，一会儿变棍子戏法，一会儿跟麦当娜玩抢棍子游戏。

10. TigerWoods and Madonna noticed a leopard tortoise crawling along the electrified fence before it was moved by staff to a safer area. Madonna was unhappy and lost her temper, snarling. To satisfy her, the tortoise was placed down by the river bank, where the cubs could continue to watch it.

虎伍兹和麦当娜看到一只豹龟沿着围网爬，工作人员捡起龟来放到别处。麦当娜见状很生气，直发脾气。豹龟只好被放在小虎们仍能看见的河床上。

11. Tigers love water. Even the water trough was not safe. Madonna used it as a bath. When she climbed out, TigerWoods attempted to climb in. But he didn't want to get his front paws wet, so he lowered himself into the water bottom first. Then he tried turning with his paws clutching the edge, but it didn't work. Splash! So TigerWoods got his front paws wet after all. He jumped straight out, but jumped straight in again, sloshing so much water out of the trough as he did so.

老虎喜欢水，连水槽都不放过。麦当娜从水槽出来后，虎伍兹又爬了进去，他不想弄湿自己的前爪，就后身先下，再笨拙地转过身来，把前爪也浸在水里，进进出出好几次，像在跳芭蕾。

12. Branches snapped and cracked in the tree. Madonna had climbed a full three metres from the ground and was still aiming higher. Was she trying to catch some shrieking monkeys? When the monkeys absconded, Madonna remained on the tree, angrily biting its branches.

树上传来了树枝折断的声音——麦当娜竟然上了树！她爬到3米多高的地方，可能是想捉住那几只尖叫的猴子吧。猴子却不知道哪里去了，但麦当娜仍执著地不肯从树上下来，坐在树叉上啃起了树枝。

Walking with TigerWoods and Madonna

A couple of weeks ago Madonna was observed with some bare patches of skin, a sign of fungal infection. We called on a wildlife vet, Joseph Van Heerden, to come and treat her. Joseph has many years of experience in treating African big cats. I was greatly impressed and relieved when he told me tigers are more like cheetahs and often react badly to anaesthetics. He had reservations about the conventional anti-fungal drugs and sedatives and came up with a different and safer alternative to treat Madonna.

Madonna reacted very well to his concoction and woke up in three minutes after the antidote was applied. Instead of giving the tigers conventional tablets and regular medicinal wash Joseph prescribed special capsules, Sporanox. These have never been used on animals before and while expensive are a lot safer.

Since Hope's death we have been trying to train TigerWoods and Madonna to get into the crush cage in case they would need medical attention. However, having seen how the older tigers were treated in the

样！她打着友好的噗噗，头蹭着国泰，表示亲热。

国泰也立即以友好相报，马上玩耍了起来。她对虎伍兹是尽可能的严厉，而对麦当娜则尽可能的温柔。3只虎以快乐的游戏、追逐结束了这个美妙的下午，国泰脸上仿佛洋溢着笑容！

与虎伍兹和麦当娜遛弯儿

两星期前，麦当娜出现掉毛，这是真菌感染的迹象。我们请了约瑟夫·范伊登兽医来看病。约瑟夫是南非顶级野生动物兽医，6个月前刚刚搬到离老虎谷仅两个半小时的金伯利。尽管约瑟夫没有医治过老虎，却有着多年丰富的医治非洲大型猫科动物的经验。来之前他还做了详细的研究，听到他说老虎如同猎豹一样对麻醉药反应强烈后，我对他由衷的敬佩，也松了口气。他对常规抗菌药和镇静剂有很大顾虑，对麦当娜使用了与众不同更安全的治疗方法。

麦当娜对镇静剂反应良好，解剂应用后3分钟她就醒了过来。约瑟夫医疗老虎的处方也与以往的兽医们不同。他开了独特的胶囊"Sporanox"，而不是常规的药片和洗剂，这些胶囊以前还没被用来治疗过动物，虽然价格高昂，但要安全得多。

cage, they refused to even walk into the half hectare camp where the cage is located, never mind get into the cage! My team were resigned to this defeat and therefore were only too happy not to have to use any medicine which requires the use of the cage.

It was, however, no small feat to give the tigers drugs in turn the next morning. Both Madonna and TigerWoods needed to be treated for fear of cross infection and they must be separated to prevent one eating the other's share. After they each got their medicine stuffed inside meat I sat with them at the fence. Then I decided to take a walk along the fence.

To my amazement, first Madonna then TigerWoods got up and followed me on their side of the fence playing with each other as they went along. I wanted to test if they were intentionally following me so I suddenly stopped in my tracks. The tigers also stopped, looking at me as if puzzled and waiting for direction. When I pretended to turn and walk back they copied me. When I 'changed' my mind and walked further along the fence towards the half hectare camp, they again followed me. When I needed to walk away from the fence to avoid some thorny plants they would wait for me on the other side. When they had to do the same they would run quickly to catch up with me. They were chuffing at me the whole time, especially after these little 'hiccups' with plants and turns. I felt so incredibly honoured to discover that these tigers were walking with me!

I wanted to see how they would react when we reached the half hectare camp which they dreaded. When I reached the half hectare gate I walked passed it. TigerWoods stopped at the gate. In order to follow me he will have to walk over into the half hectare camp through this scary gate. He hesitated. But seeing that I walked further and further away he suddenly drummed up some courage and made a little jump across the gate into the camp as if there were a huge barrier on the ground which prevented him from simply walking across!

Madonna arrived, crouched down and refused to make a move. I tried to encourage her. She stood up hesitating. Her front legs moved wanting to walk into the half hectare camp but stopped in mid-air as if there was this invisible glass door stopping her from making an entrance. I continued calling her and chuffing at her, and as if to prove to me that she would do everything for a friend, she finally took the plunge and also jumped across into the camp!

Thereafter, feeding medicine became an easy job as Madonna (almost always the first!) and TigerWoods got into the routine of walking with me, and the gate would close after one had walked into the half hectare camp to separate them. They would then each receive their portion of medicine-filled meat. They seemed looking forward to this routine as they would either be waiting at the fence when I drove up or would quickly come to join me as if they feared missing the walk every morning and afternoon. I also discovered that they were much more relaxed during the walk and loved playing at the dry river bed so I would stop there to let them rest, roll and play. On occasions when other people came along they would become hostile and tense, refusing to relax and play, sometimes even snarling as if to say this is our private tiger walking moment and intruders are not welcome.

I am so proud that I am being treated as a member of the clan by these young tigers. I am grateful for this honour, particularly since they are on their way to becoming proficient hunters, and it is an experience that I will treasure forever.

希望死后，我们团队一直想训练麦当娜和虎伍兹使用压缩治疗笼，为今后治疗的需要做准备。然而，虎伍兹和麦当娜可不傻！由于亲眼见过国泰和希望是怎么受骗，在治疗笼中受过任何种待遇，两只小虎拒绝进入安放治疗笼的半公顷营地中，更甭说踏进笼里了！我的团队只好以失败告终。所以，不需要治疗笼就能给老虎吃药让队员们很高兴。

能让虎伍兹和麦当娜吃药也不简单！为防止相互感染，两只虎都要接受治疗。我们把胶囊的药粉放进肉块中，喂药前先把他俩分开，保证每只虎都能吃上药。喂药后，我坐到围网边观察他们，然后起身想围着围网走走。

令我惊异的是，麦当娜和虎伍兹都先后站起来，隔着围网跟随着我，边走边玩耍。我想知道他们是不是故意跟着我，就做个小试验，突然在小道上停了下来。虎伍兹和麦当娜也停住脚，不知所措地看着我，不知往哪个方向去。我转过身往回走，他们也照办。当我改变主意继续向前朝半公顷营地走时，他们也跟着。当我远离围网躲进荆棘时，他们就在围网那边等我。当他们也遇到树丛必须绕开走时，便很快小跑赶上我。这期间，他们都朝我不停地打噗噗，尤其是遇到荆棘和拐弯的时候。我感觉到无比的荣幸，老虎居然在和我一起散步！

我想看看到达半公顷营地时他们有什么反应。这营地虽小，在他们眼中却最可怕，他们年长的同伴们曾被关进可怕的治疗笼，身上喷洒难闻的药水。当我到半公顷营地的大门时，我走了过去。虎伍兹停在了门前，如果要继续跟随我，他必须跨过这可怕的铁门进入半公顷营地，他犹豫不决。看着我越走越远，他突然鼓起勇气，身体一跃，穿过大门进入了半公顷营地，仿佛地上有个大门坎不能让他平步走进来似的。

麦当娜到门前就蹲伏在地，不肯再前进。我给她加油，她站起来，犹豫不决。她前爪挪动很想进入半公顷营地，但爪子停在了半空中，好像有一道看不见的玻璃门阻挡着她进入。我不停地呼唤她，朝她打噗噗。似乎向我证明她为朋友可以两肋插刀一样，她终于下定决心，使劲一跳，也进入了半公顷营地！

从那以后，喂药成了一件轻而易举的事，麦当娜（几乎总是第一！）和虎伍兹把与我遛弯当成例行公事，每在一只虎进入半公顷营地后，门就会关上，把他俩分开。然后，他们就各自得到一份儿塞有药物的食物。每天清早和下午，他们都在围网边等我开车过来，或者听到我来了就立即睡眼惺忪地走过来，生怕错过与我散步的机会。我还发现他们散步时很放松，爱在干河床上玩耍，所以每次我都在河床那儿停下，让他们休息，翻来覆去地打滚儿、玩耍。若有其他人在场，小虎们就会变得警惕、敌视，拒绝放松、玩耍，甚至冲着人吹胡子瞪眼、龇牙咧嘴的，好像在警告外人：这是属于我们老虎的散步时间，外人不受欢迎。

我为小虎们把我当作老虎部落的一员感到骄傲。我对这个荣誉无上感激，更何况他们正茁壮成长为高超的猎手。这种经历值得我为之付出一切。

4. Romance
第四章 浪漫恋爱

Tigers can be very picky when it comes to choosing a mating partner. But once a tigress sets her eye on a stud, her passion is unsurpassed by any other beast.

The combination of the world's most powerful predator and enflamed hormones generates romance of the most tempestuously dramatic kind. At a heartbeat's notice interludes of the utmost tenderness and affection give way to what look like battles of extreme ferocity.

Cathay was decidedly amorous. Armed with her most seductive weapons, perfumed scents, sensuous posturing and an innate ability to 'instruct' her paramour, she embarked on a mission of serial conquest. At first her pursuit, angling and bullying were rejected, scorned and thwarted. But her determination was unstoppable. She won her love battles, first with TigerWoods, then with 327.

But she also had a rival, Madonna. A complicated love triangle between Cathay, Madonna and TigerWoods ensued, resulting in jealousies, heart-breaks and real fights. But all ended well enough. Both Cathay and Madonna gave birth to the first generation of South China tiger cubs to be reared at the Laohu Valley Reserve.

The future of the South China tiger was perhaps not so bleak after all.

其实老虎在选择配偶上是出人意料的挑三拣四。然而，母虎一旦选定了对象，她的激情是任何动物都无法比拟的。

世界上最强大的食肉动物与炽灼的荷尔蒙组合，能产生最激情、最富戏剧性的浪漫。甜蜜的温柔会在一瞬间转换成看似凶猛的争斗。

国泰陷入了情网，她浑身散发着最诱人的魅力：芳香的气味、性感的仪态，与"使唤"情人的天赋。她展开了征服的历史使命。起初，她的追求与霸道遭到拒绝、阻挠与蔑视。但她坚定不移，终于胜利，先是赢得了虎伍兹的爱情，后又受到327的青睐。

但是她遇到了情敌——麦当娜！国泰、虎伍兹与麦当娜之间产生了复杂的三角恋爱，引发起嫉妒、伤心与搏斗。不过结局并不逊色，国泰和麦当娜都成功地生产了老虎谷的第二代虎崽。

也许，华南虎的未来并不黯淡无望！

Cathay & TigerWoods

国泰虎伍兹情深意长

1. TigerWoods was sniffing at something in the scrub when Cathay sprinted towards him. Suddenly she halted in her tracks. TigerWoods had turned to face her, flaring his shoulders so that he looked much bigger than usual. He was not pleased. Cathay's tail dropped between her legs as she sloped off. But she was in no mood to give up. As soon as TigerWoods had turned his back she snuck up and pounced on him. Standing upright on their hind legs they began pawing each other in an amicable test of strength.

虎伍兹正在闻着什么，国泰突然向他跳过去，最后一秒才站住脚。虎伍兹转过头来耸起肩膀站在那里，这姿势让他显得大了很多。国泰吃了一惊，夹着尾巴扭头就走。但她并没有放弃，虎伍兹转过身后，她又偷偷地靠近扑了上来玩闹，二虎都站立起来用爪子互相拍打，试探着对方的力量。

2. TigerWoods and Cathay went to the river and continued their game of circling each other in the water. When they came close they would gently paw one another.

虎伍兹和国泰到河边，在水里继续他们的游戏——转来转去，互相偷袭，时不时地给对方轻柔的一巴掌。

3. After being separated from TigerWoods for a while Cathay embraced him with her front paws and gently nibbled his neck. Later, TigerWoods furtively touched her rear. Cathay let out an angry roar and made to hit her suitor. According to the Chinese saying, 'A tiger's rump may not be touched.' A real fight ensued. TigerWoods was gentleman enough to lie on his back, a position of submission, though he continued clubbing Cathay with his big front paws.

分离一段后，国泰一见到虎伍兹就一把抱住他，轻柔地咬他的脖子。之后，虎伍兹悄悄地用爪子碰了一下她的屁股。老虎的屁股哪能随便摸？国泰吼叫着转过身来，两只老虎开始了一场颇为激烈的战斗。虎伍兹虽然仍然处于下风，但也毫不示弱地用力将爪子挥向国泰。

4. Cathay loved TigerWoods and would sometimes ride him. TigerWoods always let her have her way and would only gently push her off if she had no intention of getting off by herself.

国泰非常爱虎伍兹，有时骑到他身上去。虎伍兹任她骑，但国泰如果骑个没完，他只好小心翼翼地把她推开。

5. On 9 August TigerWoods met Cathay again and Cathay set about seducing him as she had 327. She crouched in front of him, baring her rear. TigerWoods, the younger of the two, at last gave in and agreed to mount her. What happened next was just like tinder meeting flame. TigerWoods mounted Cathay many times, each time changing his position. When he had finished both tigers roared, before Cathay turned to face her partner and give him a slap, perhaps to remind him who was really boss. Thus reproved, TigerWoods trotted off, his tail between his legs. When he had gone, Cathay celebrated by lying on her back and rolling right to left and back again. A successful mating had taken place.

虎伍兹又与国泰相见，国泰也像对327一样，引诱起虎伍兹来，在他面前低下身子。年龄小、一直很谨慎的虎伍兹，终于放下戒心，第一次骑跨到了国泰的身上。自此之后，两只老虎就像干柴碰烈火那样一发不可收拾，频繁骑跨起来，终于，两只老虎都发出了吼声，国泰翻过身来给了虎伍兹一巴掌，虎伍兹逃之夭夭。之后，国泰在地上翻来滚去。交配成功了！

6. It was time for Cathay to mate again so she was separated from her sons. As the gate between them opened she ran eagerly toward her lover TigerWoods. All along he had been waiting for her. They rubbed heads, snuggled up to each other, rolled together in the grass, then played a game of 'catch me if you can'.

让国泰与儿子们分开，是为了她能再继续发情交配。国泰盼望已久，大门一开不需"动员"就自觉地跑向了早在那儿急切等待她的虎伍兹。分别已久，小两口一见面就不住地拥吻、蹭头，快活地在草地上追逐打闹。国泰开心地爬到虎伍兹的背上又滑下来。

7. Cathay delighted in climbing on TigerWoods's back whenever she could. But her mood seemed to abruptly change. She left off playing and trotted over to her favourite tree, standing tall and straining for its branches. Shunned by his sweetheart, TigerWoods crept up jealously and menacingly behind her. Cathay swivelled round, roared, and chased him away.

TigerWoods had been given an ostrich the day before, but, picky as ever, he hadn't taken so much as a bite. Seeing two vultures attempt to steal it, Cathay charged at them, and then carried the ostrich carcass into the dense trees for cover.

一会儿，国泰又走到大树下，高高地站了起来，使劲够着树枝。虎伍兹见心上人不理他了，就吃醋地偷袭她。国泰吓了一跳，大吼一声翻身紧追。

头天喂虎伍兹鸵鸟，挑剔的家伙一口没动。而国泰蹲下刚吃了一点儿，两只秃鹫就飞落到鸵鸟尸体上，国泰像出膛的炮弹冲了过去，捍卫食物。为了防止其他不劳而获的小偷，她把鸵鸟叼到树林隐蔽了起来。

8. Reunited, in the golden afterglow of the sunset Cathay tiptoed toward TigerWoods. She lay down beside him and held him with her paws. TigerWoods turned over and returned Cathay's embrace, revelling in nature's beauty. Soon Cathay fell asleep, but not so her partner. His eyes wide open, he seemed to savour a moment filled with love and ease.

国泰和虎伍兹久别重逢。夕阳的余辉金光灿烂，国泰悄悄地来到虎伍兹身边躺下，紧紧地搂住他。虎伍兹也陶醉于这金色时光和爱慕之情之中，他转身将国泰抱在怀里。国泰很快便睡着了，而虎伍兹却不愿就这样睡去。他睁着双眼，享受着这不同寻常的充满爱情的气息。

Diary by Li Quan: Game of Big Cat Love

全莉日记: 大猫的爱情游戏

16 December 2007

Anyone who is 'owned' by cats understands why the term 'sex kitten' is applied to attractive and seductive women as, indeed, I can never resist a cat's seductive power. Among the many writings dedicated to cats one writer said, 'One either loves cats, or does not know them.' Let's rather hope this is so, better than believe there are actually people who are not touched by the enchanting power of cats!

Watching the game of love being played by the biggest cats of all is a great privilege for which I am truly grateful. Cathay started showing signs of oestrus two days ago and allowed TigerWoods to smell her. Yesterday afternoon, she was her most seductive self and repeatedly rubbed her head against TigerWoods, cuddling up to him, slapping him gently, nudging him playfully, and jumping over him light-footedly in an

2007年12月16日

任何被猫占有的人都会明白为何用"性感猫咪"这个词来形容性感的人类女性。要抵抗猫的妖媚非常难。众多赞美猫的作家中，有一位曾说："人要是不爱猫，就是不了解猫。"我宁可这么认为，起码这比相信确有不被猫的魅力所感动的人好。

能看见世上最大的猫科动物玩儿爱情游戏是一大特权，为此我非常感激。两天前，国泰开始有发情迹象，允许虎伍兹嗅她。昨天下午，她显示出了最诱人的一面。

她一会儿用头蹭虎伍兹，一会儿又拥抱他，轻轻地用爪子拍他，玩儿弄地推搡他，轻松一跃地跳过他，挑逗虎伍兹与她圆房。可虎伍兹刚刚摆好交配的架势，正想咬国泰的脖子时，她又一溜烟儿跑掉了。虎伍兹也不想被当成妻管严，受到冷落时，他竟到一边儿自己卧下，虽然此时国泰已在他面前摆好了圆房的姿势！

effort to tempt him to mate with her. However, when TigerWoods did try to bite the neck of Cathay in mating position she dashed off. Further, he did not want to admit he is henpecked. When he was snubbed, he would go away and lie down all by himself even when Cathay had actually positioned herself right in front of him!

I don't know what happened at night. But they seemed to be tired this morning and went to sleep even before the sun started to shine.

In the afternoon I heard the distinctive roars coming out of the 9 hectare camp and rushed over finding Cathay rolling on the ground belly up. I waited for nearly one hour before another successful mating occurred. It might be due to the heat that they were not mating as

我不晓得那晚他们有无房事，只是今天早晨看到他们显得很疲倦，太阳还没升高，就双双早休去了。

下午，听见9公顷营地传来的那种特殊吼声，我连忙赶了过去，看到国泰正肚皮朝天地打滚儿 我又等了近一个小时，才看见他们再一次成功交配。可能是天热的关系，他们交配的频率没8月份那么多。有两次他们怎么也配合不好。看到虎伍兹还忙着摆姿势国泰就溜掉时，我能感到国泰的气馁与烦恼。

自国泰产出首只虎崽，至今已有3个星期了，很可惜我们不得不人工饲养虎噜。我祈盼100天后，国泰与虎伍兹深厚的感情又结出新的果实，而且国泰这次能学会自己养儿育女。

frequently as in August. There were a couple of attempts where the couple struggled to coordinate their acts. I could feel Cathay's frustration and annoyance, as she sauntered off from under TigerWoods while he was still busy positioning himself.

It has been three weeks since Cathay gave birth to her first cub, Hulooo, whom we unfortunately had to hand rear. I pray that the deep love between her and TigerWoods will result in more fruits in one hundred days and that Cathay learns to raise her own children this time.

Madonna & Tigerwoods

虎伍兹麦当娜青梅竹马

1. One afternoon in October 2006 TigerWoods showed a behaviour trait not seen previously. Numerous times he mounted Madonna. Though he was out of position sitting only half way up her back, the rest of the mating posture was there: Madonna lying in front of TigerWoods, and he got on her back, nibbling her neck and moaning.

2006年10月一天下午，虎伍兹突然做出了新动作，无数次地爬到麦当娜身上，不过位置不正确，几乎是骑在她的腰上，但其他的动作都挺像模像样了：麦当娜卧在虎伍兹面前，虎伍兹骑在她背上，咬着她的后颈呜呜叫唤。

2. TigerWoods repeatedly tried to mount and mate with Madonna, often moaning as he did so. For Madonna, however, it was just another game. She would chuff at him, allow him to play with her for a while, then shrug him off. Sometimes she rolled on her back, to show TigerWoods her stomach, and make it impossible for him to mount her. To try his luck again, TigerWoods might sit beside Madonna and rub heads with her. If she rolled back over he would go to her rear in the hope of mounting. But if she wasn't in the mood Madonna just got up and left.

虎伍兹几次爬到麦当娜身上，做出交配的模样，还哼哼叽叽地叫，麦当娜却当是新游戏，翻过身来和他嬉闹一阵，友好的打招呼，然后离去。有时麦当娜躺倒在地，肚子对着虎伍兹不让他上去。虎伍兹又会想方设法，坐在麦当娜的身边，蹭她的头，起身绕到她身后想爬上去。麦当娜却干脆起身走了。

It is always a spectacle to watch young tigers play. Sometimes all that one can seen is a blob of stripy orange in a tumbling swirl around the bush or hurtling through the tall grass. Anyone who has watched the games and pursuits of domestic kittens will have an inkling of what big game cats can get up to. But what needs to be seen in the flesh is the raw, bouncing power of the tiger cub learning how to survive and hunt.

Being the only cub in the first litter born at Laohu Valley, and hand-reared by human parents, Hulooo had to learn he was a tiger and, when more cubs appeared, to discover the joys of playing with his younger siblings. At the first meeting, Hulooo was discernibly terrified. Coco and JenB were mischievous partners in crime when not ambushing each other.

Princess and King Henry, born unattended in the reserve's wilderness, were initially shy of human encounters. But if Princess has remained somewhat demure, not unlike Madonna her mother, King Henry has become decidedly gregarious.

Each day at the Laohu Valley Reserve the activities of each tiger is carefully monitored and recorded. Reports are then sent to the trust's advisers, and to interested scientists.

Here, for example, is an excerpt from a daily tiger diary, entitled 'Attack of the Killer Bush':

Hulooo was attacking a bush, pouncing on it and twisting himself into knots on top of it, trying to claw it with his back legs. He then walked to the other side of the water trough where the bush could not see him and stalked it, pouncing on it unawares and biting it to death. JenB and Coco were watching this whole episode but decided to play with each other just in case the insanity was contagious and they too were infected. Hulooo gave up on the bush, dipping his foot into the water trough and shaking it around, occasionally chewing on the tap. JenB had to have a look-see and Hulooo found it necessary to guard the trough from his advances by positioning himself between JenB and the water. Hulooo then returned to the bush to continue bullying it into submission. He later left and joined his brothers in sleeping off all their exertions.

幼虎的嬉戏玩耍是件奇观。通常我们所能见到的，只是在灌木丛间滚来滚去的带条纹的橙色毛团，或是闪现在深草中的鱼跃猛冲。看到过小猫互相追逐的人，可以从中窥见大型猫科动物游戏的端倪。其实，这些精力充沛、充满好奇心的小老虎们，是在学习生存与狩猎的技能。

虎噜是老虎谷降生的首胎，也是那一胎中唯一的虎崽，由人类父母一手养大。兄弟出生后，他经过重新认识自己是老虎的过程，学习与虎兄虎弟玩耍，但是他初次与小弟弟们见面时却被吓得心惊胆颤。扣子与金箍棒则是一对淘气的犯罪伙伴，不是之间相互伏击，就是一起恶搞。

小公主与亨利王在旷野中出生，起初非常害怕人类。小公主虽然至今还很腼腆，很像母亲麦当娜年轻时的性情，但亨利王却变得很爱交际。

项目的实地工作人员每天都对老虎的活动进行密切监控，并报告给项目的顾问与科学家们。以下内容摘自题为《进攻灌木杀手》的老虎日记：

"虎噜正出击一丛灌木。他猛扑过去，在上面扭动着身体，试图用后爪将它钩牢。随后，他踱到了水槽的另一边，成功躲避了灌木的跟踪视线；然后又蹑手蹑脚地接近灌木，出其不意地袭击，将其'咬于死地'。金箍棒与扣子见证了事件的全过程，决定彼此嬉戏，免得受到这种荒唐行为的传染。虎噜放弃了与灌木丛的纠缠，将脚爪浸入水槽搅和起来，还不时咬几下水龙头。金箍棒忍不住想凑过来看个究竟，虎噜赶紧保卫水槽，身体挡在了金箍棒与水槽之间。后来，虎噜返回灌木丛，继续对它恐吓威胁、使它俯首称臣，虎噜晚些时候才离开，加入弟弟们的行列呼呼大睡起来。"

Hulooo's Antics

虎噜的滑稽

Hulooo, Born on 23 November 2007

虎噜: 2007年11月23日出生

1. Hulooo was moved into the tree camp where he began exploring his new environment with exaggerated caution. He was almost five months old. On the one hand he needed to start distancing himself from his human parents, and learn how to become a proficient hunter. On the other, he was still so small that in the wild especially he continued needing the equivalent of a mother's care.

虎噜进驻了树营地，他小巧可爱，好奇地、小心翼翼地闻着地上的草，熟悉新环境。他快5个月大了。一方面需要逐渐远离人类父母，训练成为自力更生的捕猎者，另一方面又不能让他幼小的心灵因失去"父母"而受到创伤，在野外，他也还需要妈妈的照顾。

2. Hulooo was taken out of his tree camp to introduce him to the narrow stream. He charged excitedly toward the tall reeds without realising they were next to water. He jumped into the reeds and got a wetting. Startled and embarrassed, he ran back like a child who had made a naughty mistake. But when he heard his father TigerWoods chuffing at him he turned back into the reeds. The sight of the stream made his paw shrink involuntarily. He slipped, and got soaked again.

虎噜被带出营地遛弯儿，适应小河的水。他一溜烟跑出，朝着前方高高的芦苇丛冲了过去，并不知道旁边就是小溪，远远地就往芦苇上跳，一下子掉进水中，被惊得愣了一下，羞答答地好像小孩儿犯错儿似的又往回跑。听到虎伍兹对他打着招呼，他就用芦苇作隐蔽，想要偷袭虎伍兹。但对小河还是害怕，不小心脚爪一出溜，陷进了浅浅的溪水中，吓得赶紧缩了回来。

3. Hulooo could be amazingly perceptive and co-operative. One day, seeing that some staff were weeding the stony area in front of his camp, he too began pushing stones around with a front foot before grabbing a mouthful of the nearest plant which he then spat out.

虎噜有着敏锐的洞察力，一次，看到员工在附近除草，虎噜竟然也用前爪把石头推开，然后咬住一口草，拔了出来。

4. A tiger was lying where he shouldn't have been close to the breeding centre. This was Hulooo, waiting patiently. When Li Quan drove up in her truck, he jogged happily towards the vehicle. As Li jumped out, he jumped in. For several minutes he played havoc with seat covers and camera bags, sniffing and biting everything inside. How on earth had he got outside the fence, and why hadn't he bolted into the veldt?

远远地看到繁殖中心外不远的地方卧着一只虎！定睛一看，老天，是虎噜！他在营地外面的路上耐心地等着！见到我的车，他兴高采烈地跑过来。天知道他是怎样翻过了铁丝网！好笑的是他虽然可以"逃亡"，却竟然守在营地门的外面站岗！趁我下车时，虎噜居然跳进车里，接下来一片混乱，虎噜爬前爬后就是不下车，一会儿扯坐垫，一会儿咬相机包。

5. Hulooo was let loose on the open grassland and in no time caught a fowl. He let it go only to chase it again. The fowl was not dumb either and kept crawling into the dense bushes to take shelter. Hulooo showed his versatility by diving in amongst the branches to fish the fowl out.

虎噜来到外面开阔的草地上，一小会儿就抓住了一只野鸡，不过并未致其于死地，而是欲擒故纵，反反复复。珍珠鸡也不逊色，不断往大树旁边浓密的矮树丛里钻着躲避。但虎噜竟然钻进树丛里把它掏了出来。

6. Eating a blesbok proved to be a challenge for Hulooo. However hard he tried he was unable to break through its skin. He tried everything from standing on it, rolling it over and twisting its legs around, all to no avail. He even hugged the blesbok around the neck and whispered sweet nothings into its shrivelled ear.

大羚羊是个挑战，努力了多次，虎噜还是怎么也无法撕咬开它的皮，把它翻过来、拧它的腿也没用，但他还在不懈地努力着，居然紧紧搂住羚羊脖子，对着猎物的耳朵"甜言蜜语"起来。

194

Cathay's Sons: JenB and Coco
国泰的小儿子们金箍棒与扣子

JenB & Coco, Born on 30 March 2008

金箍棒和扣子：　2008年3月30日出生

stamina, and there was another near encounter. The hunt continued as the sun disappeared completely and judging from the constant moving shadows of the blesbok, I could guess the tigers were still hard at work. I bet they would make another kill tonight!

6 August 2009 - Battle Between JenB Brothers and Blesbok

Having not witnessed any kills, never mind a dramatic hunt for years since 2006, I was richly rewarded today. My friend Nicky flew in with her family from Johannesburg this morning, bringing my husband Stuart with them.

Inside the hunting camp, I was just wondering which direction to go to find Hulooo and brothers when I saw blesbok running madly in the hilly area, and shadows of running tigers here and there. We quickly drove towards the west, where the blesbok were scattered. We watched two tigers disappear behind the trees not far ahead and followed them down to the river area. In the distance against the sun, there was a silhouette of a blesbok, and a tiger!

My goodness! The blesbok was charging the tiger! My heart raced as I anticipated the third fantastic battle I ever witnessed between tiger and blesbok during the history of our rewilding.

JenB tried, almost half-heartedly, to bite the blebok's neck but the blesbok kept avoiding it. However, standing in the pool of water, the blesbok kept slipping each time it moved, giving the tiger advantage to tackle him. After a few rounds, JenB, a bit confused what to do with a slipping blesbok, finally grabbed it by its neck, somewhat softly, and

开车进入狩猎营地,我正考虑往哪个方向去才能找到虎噜弟兄时,突然看到羚羊们在小山那边疯狂乱跑,老虎们的身影也时隐时现。我们迅速朝西驶去,羚羊漫山遍野,四处奔命。看到不远处两只老虎消失在一棵树后,我驱车跟向河边,逆着阳光,我猛然看到不远处一只羚羊和一只老虎的轮廓。

老天!!!那只羚羊正在攻击老虎!我心情无比激动,自野化训练以来,这将是我第三次目睹这种激动人心的狩猎场面!

金箍棒仿佛三心二意似的企图咬住羚羊的脖子,但都被它躲开了。可是站在河床水中,羚羊总是打滑,给了老虎制服它的有利条件。可金箍棒看到羚羊打滑有点不知所措。几个回合后,他终于小心翼翼地咬住了羚羊的脖子,把它拖到深草后边去了!羚羊的呻吟声慢慢消失了,接下来的几分钟陷入了死一般的沉寂。

我为金箍棒感到骄傲,我们还以为他很不争气呢。

真是好事成双,因为稍后我们发现河边山岩上还有一只羚羊尸体,扣子和金箍棒正轮流试图把它拖下来。经过不懈的努力,扣子终于巧妙地将腿卡在岩缝里的羚羊拖了出来。

三兄弟的进步是显而易见的,现在数数羚羊还剩21只,说明昨晚到现在他们又捕获5只,从周日进驻打猎营地以来,他们4天内总共猎捕了8只白面羚羊。

2009年8月9日 从追捕开始到狩猎成功结束

5天成功狩猎10次,我们推断他们不会再轻易狩猎成功了,因为年轻易猎的羚羊都葬身虎爪。我和来客赫克托·玛浩米博士(南非国家保护区总生态学家兼我基金会项目顾问)及其儿子准备早点离开狩猎营地。

就在此刻,赫克托发现营地远方一只虎跑了起来,原来是两个虎兄虎弟正在跟踪一只羚羊。

dragged it behind the tall grass! The moaning of the blebok died eventually, followed by a few moments of dead silence.

I am so proud of JenB, whom we thought had been the lame duck.

Another surprise laid in store, as we discovered a second kill lying on the hilly rocks of the river bank. Coco and JenB tried in turn to drag it down to a more accommodating area. Only after much effort, Coco managed, in a clever move, to dislodge the blesbok whose legs were caught in the rocks.

The progress of these young tigers were remarkable. We counted 21 blesbok now remaining which means they made five kills since last night, bring the total number of kills to eight in the last four days since they entered the Hunting Camp.

9 August 2009 - From the Beginning of the Chase to the End of the Kill

Having made ten kills in five days, we concluded there weren't going to be any more kills that easily anymore, as the young buck seemed to have all been hunted out. After a short drive, I and my guests, Dr. Hector Magome (Chief Ecologist of South African Parks and advisor to the charity) and his son were ready to leave the Hunting Camp.

Just then, Hector saw a tiger running in the far distance. It turned out two brothers were trailing a buck.

The three tigers now positioned themselves in different parts of the camp, forming a line, pushing the blesbok into the open grassy area. Hulooo made an attack splitting the buck into two groups, all fleeing for life.

Suddenly Hector cried out: 'they are making a kill! Look, look!' In the far distance, Coco and JenB were attacking one group of blesbok, scattering them. Then both tigers focused on one individual and within blink of an eye, they brought it down. What incredible coordination! While I was driving towards them, we saw the blesbok got up, trying to escape from the tiger's grasp, but was tackled and brought down again. By the time we arrived at the scene, the blesbok was still fighting for its life.

The blesbok made repeated attempts to escape but the tigers nailed it down each time. But after three or four rounds, Coco finally delivered a deadly bite on its neck...

Hours after the hunt, an excited Hector remarked, 'Li, I must thank you. In my whole 25 years as a field ecologist, I have never seen what I saw today, a hunt from the beginning of the chase to the end of the kill!'

10 August 2009 - War in the River

All three brothers were in the center of the camp heading with a purpose towards the River area, undisturbed by our arrival. In the distance, we saw a lot of commotion with blesbok running all over the hill, tailed by tigers everywhere. One tiger, which turned out to be JenB, brought down a blesbok but it managed to turn over and get up running away again, while JenB was briefly distracted by the other moving game. JenB quickly followed his prey.

By the time we arrived to investigate, JenB had already paralyzed the buck, at about the same spot where he made a kill four days ago, and was trotting away trying to catch more game. Almost like a movie repeating itself, the antelope was struggling to get up but unable to, partly because it was seriously injured and partly because it was slippery. But its struggle caught the attention of JenB again, who ran back trying to deliver the final bite. JenB thought he had finished it off and again ran off, but was once more drawn back by the noise made by

现在，3只虎各就各位分布在营地不同的位置，形成一条线，将羚羊往开阔的草地推赶。虎噜发起进攻将羚羊群分成了两组，羚羊们各自奔命。

突然，赫克托大叫起来："他们抓到了！！看呀，看！"远处扣子和金箍棒正在进攻，羚羊四分八散。之后，两兄弟都集中精力瞄准其中一只发动袭击，一眨眼，羚羊被摁倒了。多么天衣无缝的协作啊！我迅速驶向现场。那羚羊站起身想逃离虎掌，却再次被按倒在地。我们到达时，那白面羚羊还在挣扎。

羚羊几次三番逃跑，每次都又被老虎摁在地上。三四个回合后，扣子终于给出致命的一击，死死地咬住了它的脖子。我打开前灯打算靠近点去看，扣子拖着羚羊躲开了，仿佛那只是个玩具，也仿佛我们要抢走他的战利品一样。

狩猎结束都好几个小时了，赫克托还兴奋不已地说："全莉，我必须得谢谢你，我从事实地生态保护25年了，还没见过今天这样的从追捕开始到狩猎成功结束的捕猎场景！"

2009年8月10日 水中大战

我们的到来，并没转移正朝小河方向进军的三兄弟的注意力。不久我们看到远处羚羊一片混乱，到处被老虎尾随着。一只虎，原来是金箍棒，大步追到了一只羚羊，把它绊倒，趁金箍棒的注意力暂时分散时，它又一翻身站起来跑开了，金箍棒马上穷追不舍。

还是在4天前那次猎杀羚羊的地方，我们追上金箍棒，羚羊已经受伤，金箍棒正要跑开再去捕猎。就像电影重演一样，河床又湿又滑，羚羊受伤太重，挣扎着想站也站不起来。它的挣扎引起了金箍棒的再次注意，返回身来打算给它致命一击。金箍棒以为猎物终于被咬死了，就又跑开了，但是羚羊挣扎弄出的声响又把他拖了回来。几个回合后，他再也抵制不了河那头传来的呻吟声的诱惑了，我们也听到了折腾声，之前也看到扣子与虎噜追踪猎物消失在那边。

我们驾车来到了河流的另一头，穿过浓密的树枝，隐约看到扣子在及身的河水里正与一只羚羊战斗。目睹这场捕猎过程的翰因后来告诉我，那只羚羊在虎噜和扣子的追杀中掉进河里。羚羊挣扎着想离开河水，摆脱老虎的魔爪，但是羊与虎都使不上劲儿。扣子几次咬住羚羊的脖子，但都被它甩掉了。金箍棒的到来也帮不上忙，而虎噜就趴在几米远的地方，安心让兄弟处理这个棘手的问题。好一段时间后，我们看到扣子急中生智竟然坐到了羚羊的身上，把它压到水中窒息，直到它停止挣扎后才死死咬住了它的脖子。这时令扣子不耐烦的是金箍棒却帮了倒忙，居然咬住羚羊屁股使劲儿往岸上拖。但是，作为优秀猎手的扣子拒绝松口，直到确定羚羊死了才放松，整整咬了有10分钟左右。

看到这只死了，金箍棒冲回他刚才猎捕的那只羚羊前。试探几次后，他总算拿定主意咬住了猎物的喉咙，咬的时间不短，足有6~7分钟，有可能是从兄弟那儿学来的招儿吧。

短短一周多三兄弟猎杀了14只羚羊，饿不着了。我们决定偷一两只给其他的老虎，避免浪费。虎虎兄弟们好像怀疑我心怀叵测，当我们将河里的那只羚羊刚刚拖出来时，他们就赶到了。我们不得不放弃了这只。

我们把金箍棒捕杀的那只猎物装上了我的皮卡，在不被他们抓住前，我们这些"偷羊贼"迅速地离开了狩猎营地。

三兄弟好像知道我的卡车就是偷窃他们劳动果实的罪魁祸首，第二天下午他们谨慎地守着狩猎营地的大门，使我们不得进入。好几天都没给我们好脸色看。而且他们只在夜间狩猎，仿佛在展示抗拒"窃羊贼"的新策略。我8月14日离开老虎谷时，虎噜兄弟们在过去的12天里，一共猎捕了17只白面羚羊。

他们不断地打破我们的预言。

the struggling buck. After a few rounds, he could not resist the sound of commotion coming from the other end of the river, where Coco and Hulooo had pursued other blesbok, and where we had also heard distress moaning earlier.

We drove to the other end of the river as well and through the dense tree branches, we could see Coco tackling another blesbok in the river water which reached to his belly. Hein, who witnessed the hunt, later told me the blesbok had fallen into the river while being pursued. The antelope was struggling to get out of the river and out of the tiger's grasp, but neither seemed to have much leverage. Coco made repeated attempt to deliver a deadly neck bite but the blesbok always managed to shake him off. JenB who has arrived on the scene could not seem to help, while Hulooo just rested a couple of meters away letting his brother dealing with this thorny issue. After quite some time, Coco succeeded in sitting on top of the blesbok, cleverly, suffocating it under the water. When it finally stopped struggling, Coco wrapped his mouth around its neck, holding the bite, but his brother JenB attempted to drag the blesbok out of the water from its rear end, much to Coco's annoyance. However, Coco being the good hunter he is refused to let go, until he was certain he had

finished his kill off completely, which lasted for nearly 10 minutes.

Immediately after, JenB rushed back to his first kill. After a few rounds of uncertain attempt, he finally managed to hold his bite around the antelope's throat for a long time as well, as if getting the tip from his brother.

The young tigers have made 14 kills in a little over a week and they are certainly not going hungry. We decided to steal one or both blesbok to give to the other tigers. The brothers seemed to be suspicious of our intentions and ran over towards us just after Vivienne had dragged the one out from inside the river. We had to abandon this one.

We barely loaded the one killed by JenB onto my truck before the brothers approached, and we hurried away to get out of the hunting camp before they could catch up with us blesbok thieves.

For the next few days, the tigers were vigilant towards us, making kills only at night, as if to outsmart thieves. When I left Laohu on 14 August, Hulooo and brothers had hunted 17 antelopes in the past 12 days.

They keep on defying our predictions.

Concluding Words: Wild Return to China
结束语 重返祖国山野

The objective of Save China's Tigers and its Laohu Valley Reserve has always been twofold: to preserve a magnificent but threatened tiger, and to restore it to its rightful position in China itself, as King of the Mountains and of the Hundred Beasts. Only when its lost Chinese habitat has been recovered will it be possible to speak of success. Then, and only then, will the South China tiger once more do what it always did best, and govern its wild kingdom for generations to come.

Because of the pressures of a rapidly expanding modern human civilisation, any such habitat or habitats will need protection, which means the active participation of government. To that end, Save China's Tigers has been working closely with the relevant authorities in China from the beginning. It has enlisted the services of many experts. It has sponsored important ecological and economic feasibility studies identifying potential pilot reserve sites and it has contributed to conservation capacity building.

In 2006, the State Forestry Administration of China approved two locations as Chinese Tiger Pilot Reserve candidate sites. Since then, Save China's Tigers' team, in conjunction with various Chinese government departments, has worked around the clock to secure the necessary funding -- through state grants, investment, and also private donors. In 2010, the Chinese government approved an interim site destined for the re-adaptation process of the rewilded South China Tigers upon their return to China.

The Chinese year of the tiger 2010 came and went. 2011 saw the birth of six more second generation tigers at Laohu Valley, sired by none other than the tiger many believed would not be able to breed - Mr. "327", bringing the total number of South China tigers born into our project to eleven. Although 327, the number under which he was registered in the studbook kept by the Zoological Society of China, lost his life in his fight to become the supreme king at Laohu Valley Reserve, his legacy lives on and he has proven that a natural environment for the tigers can do wonders for them. His death ironically proves the success of our project to rewild zoo-bred tigers.

With fourteen tigers under our care and counting, we have increasing pressure to find them a large and natural home back in China very soon. We promised the young tigers at Laohu Valley that they will return to their former habitat soon, because that is the best and perhaps only chance of their long-term survival. Long revered by the Chinese people, though also of late sadly neglected by us, the South China Tiger seeks, and deserves, rehabilitation in its native land.

Time will not wait.

华南虎野化重引进项目的目标，不仅是拯救气壮山河的中国虎，也是向山林之王赎罪，重焕他昔日统领"百兽"的光辉。只有将中国虎遗失的栖息地恢复，中国虎才能长久生存下去，才能算得上成功。那时，也只有到那时，华南虎才能发挥他在生态平衡方面应有的旗舰作用。

由于迅速扩展的现代人类文明的压力，任何这种栖息地都需要受到保护，这就意味着需要政府的积极参与。

因此，从拯救中国虎国际基金会项目启动开始，我们就一直与中国政府相关部门合作，共同推动这件中国人所应承担的大业。基金会聘请了众多专家进行华南虎的野外考察；对候选保留地作出了生态与经济可行性的研究；而且对中国的野生动物工作人员进行了培训。2006年，国家林业局批准两个地区为中国虎试点保留地候选区。自此，拯救中国虎团队就夜以继日地为项目所需资金而奔波，连同国家各级政府部门寻求企业赞助、个人投资，以及国家拨款等等。2010年，中国政府批准了野化华南虎回国后进行重新适应的过渡区，并正在对三个作为华南虎重引进保护区的候选地进行评估。

2010年的虎年一晃即逝。2011年老虎谷又见证了6只二代虎的出生。他们都来自同一个父亲，一个几乎被所有人（包括兽医专家在内）断论无法生育的公虎——"327"！这些小虎的到来使在南非老虎谷诞生的华南虎总数达到11只。327，这个以中国动物园协会华南虎谱系注册号命名的公虎，尽管在争夺老虎谷保护区王者地位的斗争中失去了生命，但将流芳百世。他也向世人证明了自然环境能使老虎繁衍产生奇迹。他的死也从反面证明了我们项目在野化动物园出生的老虎方面的成功。

现在有14头老虎正处在我们的精心照顾下，他们还将继续繁衍壮大。我们的经济压力也愈加沉重，我们为他们在中国的大自然寻找未来的家园的责任也愈加艰巨、紧迫。

我们向老虎谷出生的第二代华南虎们作出了衷心的承诺，希望能早日带他们回娘家，这不仅是最理想的选择，也可能是他们长久生存下去的唯一机会。中国虎自始以来受中华民族的崇敬，但不幸的是在近代也受到我们的残害。他需要我们中国人重新给他一个家！

时不我待！

Save China's Tiger's Major Milestones
拯救中国虎大事记

1999: Li Quan formulated the Chinese Tiger Conservation Model after observing and analysing how wildlife conservation and eco-tourism have succeeded in Africa and then started seeking support for the South China Tiger internationally.

October 2000: Quan established the first international charity dedicated to wildlife conservation in China, Save China's Tigers, first in UK and subsequently in the US in 2002 and Hong Kong in 2003.

2001-2002: Quan initiated facilitated and co-funded Dr. Ron Tilson's survey team from Minnesota Zoo to search for wild South China tigers in China.

August 2001: Quan proposed to China to use South African expertise to help the Tiger Reintroduction project.

November 2001: Quan invited the first South African Team to visit China.

May 2002: Save China's Tigers donated infra-red cameras and equipment to the State Forestry Administration of China to continue monitoring the possible existence of wild Chinese tigers in areas such as Hupingshan and Taoyuan in Hunan Province.

2002: Stuart Bray, Quan's husband, acquired 33,000 hectares of land in South Africa to lay a platform for the South China Tiger Re-wilding program.

2002: Quan fine-tuned the Chinese Tiger Conservation Model - saving and protecting wildlife through eco-tourism by using the Chinese Tiger as a flagship and combining wildlife conservation with Chinese culture and heritage to make conservation sustainable in the long term.

November 2002: Tilson made a controversial declaration regarding the extinction of wild Chinese tigers.

November 2002: Save China's Tigers set up Chinese Tigers South African Trust as the operational arm for the Chinese Tiger Re-wilding and Reintroduction Project.

November 2002: Save China's Tigers and the Trust signed an historic agreement with the Wildlife R&D Centre of the State Forestry Administration of China on the reintroduction of the Chinese tigers into the wild.

2003: Quan recruited South Africa's top scientists and conservationists, Petri Viljoen, Jeremy Anderson, Marc Stalmans, Richard Davis, Gus Van Dyk, Hector Magome and others to aid in the project.

July 2003: Save China's Tigers and Cathay Pacific Airways signed a sponsorship agreement for six years.

September 2003: Two Chinese tiger cubs, Cathay and Hope, embarked on their historic journey to South Africa for Re-wilding Training.

November 2003: Save China's Tigers sent the first South African expert team to China for an ecological survey of ten proposed candidate sites for the Chinese Tiger Pilot Reserves in four provinces.

February 2004: A second South African expert team composed of resource economists and government conservation officials was sent to China to survey the top two candidate sites in Jiangxi and Hunan provinces.

July 2004: Cathay and Hope captured and ate their first African antelope.

29 October 2004: Another two South China tiger cubs from Shanghai Zoo, TigerWoods and Madonna left Beijing for South Africa. In January 2005 they caught their first guinea fowl at Laohu Valley Reserve.

12 March 2005: Cathay and Hope were radio-collared in preparation to move into a 600 hectare camp from the 62 hectare hunting camp.

May 2005: The book, Tang the Tiger Cub, by Claudine Kolle was published and aimed at children between the ages of 4 and 10.

May 2005: TigerWoods and Madonna caught their first blesbok for lunch.

20 August 2005: Hope sadly died of pneumonia and heart failure.

18-19 December 2005: An international workshop was held in Beijing on the Chinese Tiger Reintroduction project and major international organizations including the IUCN Cat Specialist Group attended. The Beijing proposal to hasten the progress of the Chinese Tiger Reintroduction Project was issued.

February 2006: TigerWoods and Madonna were released into the 42 hectare hunting camp with great success.

April 2006: The State Forestry Administration of China approved two candidate sites for the Chinese Tiger Pilot Reserve.

1999年，全莉女士在观察了野生动物保护和生态旅游是怎样在非洲成功运作之后，构思出了中国虎保护模式。并开始为华南虎寻求国际上的帮助。

2000年10月，全莉在英国成立了国际上首家致力于中国野生动物保护的慈善组织——拯救中国虎国际基金会，并先后于2002年在美国、2003年在香港注册。

2001年到2002年，拯救中国虎国际基金会倡导并参与资助了明尼苏达动物园帝尔森博士的考察队伍参加中国政府正在进行的对野外华南虎种群的考察。

2001年8月，全莉建议中国聘用南非专家来帮助中国的中国虎重引入项目。

2001年11月，全莉邀请了第一批南非专家团访问中国。

2002年5月，拯救中国虎国际基金会向国家林业局捐赠了远红外摄像机和摄像设备，用于在湖南壶瓶山、桃源等地继续探测野外华南虎种群的存在。

2002年，全莉的丈夫博锐先生，在南非购买了3万公顷的土地,作为华南虎野化项目的基地，为华南虎的重引入奠定了基础。

2002年，全莉进一步深化了中国虎保护模式，以中国虎为旗舰，结合中国虎文化遗产通过开展生态旅游，使野生动物保护可持续发展。

2002年11月，帝尔森宣布中国虎在野外灭绝，引起巨大争议。

2002年11月，拯救中国虎国际基金会在南非成立了中国虎南非项目中心，作为中国虎野化重引入项目的执行机构。

2002年11月，拯救中国虎国际基金会和中国虎南非项目中心与国家林业局野生动植物研发中心签署了历史性的协议，将野化成功的华南虎重引入中国的野外。

2003年，全莉建立了南非顶尖科学家和自然保护学者组成的执行团队，包括：Petri Viljoen、Jeremy Anderson、Marc Stalmans、Richard Davis、Gus Van Dyk 和Hector Magome等。

2003年7月，拯救中国虎国际基金会和国泰航空公司签订了长达6年的赞助协议。

2003年9月，两只小华南虎国泰和希望踏上了前往南非历史性旅程，进行野化训练。

2003年11月，拯救中国虎国际基金会派出了首批由生态学家组成的南非专家组对中国4个省的10个候选区进行生态考察，初选中国虎先锋保留地。

2004年2月，第二支由野生动物经济学家和政府保护官员组成的南非专家考察队对位于江西和湖南的前两名候选保留地进行考察。

2004年7月，国泰和希望首次抓捕到了非洲野羚羊。

2004年10月29日，第二批来自上海动物园的华南虎惠虎伍兹和麦当娜离开北京前往南非，2005年1月，他们在老虎谷保护区中国虎野化基地捕到第一只珍珠鸡作为晚餐。

2005年3月12日，中国虎国泰和希望戴上了无线跟踪项圈，准备从62公顷的野化营地进入600公顷的营地。

2005年5月，《唐小虎》一书出版，用于教育4至6岁的儿童。

2005年5月，虎伍兹和麦当娜捕获第一只羚羊作为午餐。

2005年8月20日，希望不幸死于肺炎和心脏衰竭。

2005年12月18、19日，中国虎重引进项目国际研讨会在北京召开，包括IUCN猫科动物专家组在内的国际组织都出席了会议，研讨会就加速中国虎重引进项目发表了北京倡议。

2006年2月，虎伍兹和麦当娜进入42公顷的狩猎营地，并取得成功。

2006年4月，中国国家林业局正式批准两个候选地作为中国虎先锋保留地。

2006年6月，拯救中国虎在香港发起了成龙老虎脸的老虎关注活动。

2006年11月,全莉女士受邀在皇家地理协会香港分会演讲。

June 2006: Save China's Tigers launched the Jackie Chan Tiger Face Awareness campaign in Hong Kong with the actor as our ambassador.

November 2006: Quan was invited to lecture at the Royal Geographic Society's Hong Kong branch.

April 2007: The David Tang Tiger Breeding Centre was completed at Laohu Valley Reserve.

23 April 2007: South China tiger with Studbook registration number 327 was relocated from Suzhou Zoo to South Africa to start the breeding program.

23 November 2007: Cathay gave birth to her first cub, Hulooo, and he was the first South China tiger ever born outside of China. It was an important milestone not only for Save China's Tigers' ambitious undertaking, but also an unprecedented achievement in tiger conservation history.

30 March 2008: Cathay successfully gave birth to her second litter of two cubs at Laohu Valley Reserve.

April 12, 2008: Madonna gave birth for the first time to two cubs at Laohu Valley Reserve. One died in birth and the other died seven days later at Lory Park due to bacterial infection.

18 August 2008: Madonna gave birth to a second litter, a boy and a girl. She has reared them on her own in the natural environment. This is another milestone in tiger conservation history.

March 4, 2009: Hulooo, Coco and JenB (Cathay's cubs) started hunting training.

November 2009: Madonna and her one-year-old cubs had made thirty kills in one period of under two months of hunting training. Adolescent Hulooo & brothers, though separated from mother Cathay at an early age, made twenty-one kills in eighteen days.

Dec 16, 2009: the first cub sired by 327 was born, but was unfortunately taken by a predator on its seventh day. This highlighted the danger of living in the wild, particularly for young cats.

Jan 2010: South China Tiger Diary was published by China Literature Press in Beijing. The book was compiled from Li Quan and SCT team's tiger monitoring diaries.

Jan 2010: SCT launched the "Last South China Tiger" body painting by artist Craig Tracy to raise awareness.

Feb 2010: SCT launched Tiger-Tram campaign in HK to celebrate the Chinese Year of the Tiger and raise awareness.

July 2010: Dr. Gary Koehler and Petri Viljoen visited the proposed interim site - Meihuashan, for the return of the first South China Tigers from South Africa to help design the next steps needed.

Sept 12 to 16, 2010: SCT held its first and extremely crucial Rewilding scientific workshop attended by some of the world's foremost wild cat biologists who evaluated the results of SCT rewilding project. They not only unanimously regarded the project as successful but viewed it as a potential model for future big cat conservation.

Oct 13 2010: SCT celebrated its ten year anniversary and launched a photo documentary book "Rewilded - Saving the South China Tiger", which chronicled the fight for survival by the South China Tigers under rewilding training at Laohu Valley Reserve in South Africa.

Jan 2011: SCT scientific advisors Petri Viljoen and Dr. David Smith visited several new candidate tiger reintroduction sites in China, including Hubei, Jiangxi and Fujian provinces.

Jan 31st, 2011, a second cub sired by the tiger 327 was given birth by Cathay. Named Huwaa, she was taken for hand-rearing on Feb 2nd and was returned to Laohu Valley on May 14, 2011. She was reunited with mom Cathay on May 21st, a successful move that defied traditional wisdom.

March 2011, Save China's Tigers Australia was approved by the Australian government.

May 17th 2011, two second generation tigers born at Laohu Valley, JenB and Coco were fitted with GPS/Radio collars for study of their hunting behaviour in the 100 ha camp.

June 2011, Save China's Tigers Fund was approved by the Chinese government and established under the umbrella China Green Carbon Foundation. This status allows SCT to legally conduct fund raising activities in China.

July 20th 2011, Cathay gave birth to two more male tiger cubs sired again by 327. He should be so proud now. The total number of South China Tigers at Laohu Valley now reaches 12.

Aug 2011: Save China's Tigers presented its tiger rewilding results at the annual conference of Society for Ecological Restoration in Mexico.

Sept 17, 2011, 327 - the only tiger who came from a Chinese zoo that didn't learn to hunt due to his mature age, went through an electrified fenced gate to challenge another mature male who instead killed him. In a perverse way this accident shows that the rewilding project has proven to be a success.

Oct 9th, 2011, Madonna gave birth to our first litter of triplet sired by 327, and ALL females. The total number of South China Tigers at Laohu Valley now reaches fourteen.

Dec 2011, Save China's Tigers presented its tiger rewilding results at the annual conference of Society for Conservation Biology (ICCB) in New Zealand.

And the Story Continues…

2007年4月，邓永锵中国虎繁育中心在老虎谷完工。

2007年4月23日，华南虎327从苏州动物园运往南非，开展繁殖计划。

2007年11月23日，国泰的首只华南虎幼崽在南非降生，这不仅是拯救中国虎项目的第一次，也是华南虎首次在国外降生。这是拯救中国虎项目的里程碑，也是老虎保护史上的重要成就。

2008年3月30日，国泰顺利产下第二胎两只幼崽，并自己抚养，无需人工干预。

2008年4月12日，华南虎麦当娜在老虎谷首次产下二崽，一只幼崽难产夭折，另只幼崽在出生7天后由于细菌感染在罗园公园死亡。

2008年8月18日，麦当娜第二次分娩，产下一公一母两只小仔。她在完全自然环境下抚养幼虎，就像野生母虎一样。这是老虎拯救历史上的又一个里程碑。

2009年11月止，麦当娜及其一岁的幼崽们在一次不到2个月的野训时间内捕获了30只猎物。尽管与母亲小时就分离，虎噜及其兄弟们则在18天内捕获了21只猎物。导致水平差异的原因，可能是麦当娜以示范方式教授经验比虎噜兄弟们自己通过实践学习需要更多的时间和精力。

2009年12月16日，327的第一只幼崽出生了。可惜的是第七天虎仔被其他动物叼走了。这个事故也说明了野外生活充满了不可预料的风险，尤其是对年幼的动物来说。

2010年1月，人民文学出版社出版了《华南虎日志》。该书由全莉与拯救中国虎团队的老虎监测日记筛选编写。

2010年1月，拯救中国虎发布了艺术家克雷格·特雷西制作的"最后的华南虎"人体彩绘，提高人们的野保意识。此举获得热烈反响，但遗憾的是在发布一个月后被YouTube网络视频网站限制浏览。

2010年2月，拯救中国虎在香港发起老虎电车宣传活动，庆祝中国虎年以及呼唤人们的关注。

2010年7月，盖瑞·科勒博士和帕蒂·维伦到梅花山考察，帮助中国为首批华南虎从南非的返回制定计划。

2010年9月12日至16日，拯救中国虎首次举办及其重要的野化科学研讨会，世界上名高众望的野生猫科动物学家们齐聚一堂，评估拯救中国虎的野化项目结果。他们不仅一致认为该项目的成功，并声称这有可能成为未来大型猫科动物的保护模式。

2010年10月13日，拯救中国虎庆祝成立10周年，同时发行纪实摄影集《野化拯救华南虎》，此书通过400张左右的图片讲述了华南虎在南非老虎谷野化训练，为重振王者之风而努力奋斗的故事。

2011年1月，拯救中国虎科学顾问皮特·维伦先生和大卫·史密斯博士对中国的几个新选的华南虎重引入候选地进行考察，包括湖北、江西和福建等地。

2011年1月31日，327号虎的第二个幼崽再由国泰所生，取名虎娲。她于2月2日被取出进行人工喂养，并于2011年5月14日带回老虎谷。她与妈妈国泰于5月21日团聚，这是一个证明传统智慧有误的成功实践。

2011年3月，拯救中国虎国际基金会得到澳大利亚政府的正式批准。

2011年5月17日，两头出生在老虎谷的二代虎，金箍棒和扣子被装上卫星定位GPS颈圈，对他们在100公顷营地的狩猎行为进行监测和研究。第一头被上GPS颈圈的羚羊在一周内由金箍棒捕杀。第二头被上颈圈的羚羊也在被释放后的几天内被老虎猎捕。这些狩猎结果显示了二代虎们惊人的狩猎能力。

2011年6月，中国虎拯救基金在中国绿碳基金下正式成立，得到在中国合法开展募捐活动的资格。

2011年7月20日，国泰再次生育了327的两只雄性幼仔。327现在可以趾高气扬了。老虎谷保护区华南虎的总数现在达到12头。

2011年8月，拯救中国虎国际基金会在墨西哥举行的国际生态恢复协会的年度大会上介绍了华南虎野化的结果。

2011年9月17日，来自中国动物园的唯一因为年长未经野化训练的公虎327，冲过带电铁丝的大铁门进入毗邻的老虎营地攻击另外一头公虎，但反被对方杀死。从某种意义上来说，这个事故从反面证明了野化项目的成功。

2011年10月9日，麦当娜再次生育了327的3只幼仔。不仅是我们首次的3仔胎，而且三头全都是珍贵的母虎。使老虎谷保护区华南虎的总数达到14头。

2011年12月，拯救中国虎国际基金会在新西兰举行的国际保护生物学协会的年度大会上介绍了华南虎野化的结果。

我们的故事还在继续着…

Acknowledgements

The founding of Save China's Tigers (SCT) in 2000 and its subsequent undertakings would have been impossible without the generous support of many people, including volunteers, from all walks of life around the globe, and quite a few of whom I have never met. I have also dragooned many friends into the cause and I am grateful for their continuing help.

First and foremost I would like to thank my husband Stuart Bray without whose personal and full financial support this project would have been a non-starter. His continued backing through his team at Conservation Finance International has enabled SCT to make many bold moves. We have also benefited from the following organizations either through sponsorship or from pro-bono services: Abercrombie & Kent, Asian Tigers Group, Bateleurs Flying for the Environment, Cathay Pacific Airways, Eight Partnership, The Excelsior Hotel, Goldman Sachs, Green Dragon Fund, Highveld Taxidermists, JCDecaux, Leicester Tigers, Life of Circle, Milbank Tweed Hadley & McCloy LLP, Patrick Mavros, Shanghai Tang, Tiger Wheel & Tyre, Touchmedia and Zoomania.

Gracious donations from individuals such as Jenifer Bone, Colin Williams, Severin Wunderman and many others formed a backbone to the project. Appeal Patron Sir David Tang's Tiger Fundraiser at his China Tang Restaurant was particularly memorable. A special mention should also go to Jane Lovell who led her pupils at King Henry VIII Preparatory School in Conventry in England to raise funds for SCT.

The generosity and dedication of our Appeal Patrons has been crucial in generating public interest. Over the years we were able to enlist: Jackie Chan, Kaige Chen, Maestro Christoph Eschenbach, Wen Jiang, Lang Lang, Lewis Moody, Nick Rhodes, Michelle Yeoh, Coco Jiang Yi, Da Ying and Zhongxiang Zhao. We were all very moved when Jackie Chan showed up on his own without any fuss at our launch of his Stripy Tiger Face campaign.

I am grateful for the vision and courageous initiatives that accompanied those brave individuals who placed the tigers' interests ahead of their career and a special thank-you goes to Weisheng Wang and his team at the State Forestry Administration and the Chinese Academy of Forestry Science. The Cultural Division of the Chinese Embassy in London provided us with the venue to launch the charity. Former Chinese Ambassador Guijin Liu was especially kind to the tigers and welcomed their arrival each time during his assignment in South Africa. I also want to thank the South African government's support, especially the Northern Cape Conservation Department led by Albert Mabunda, former South African Minister of Culture Dr. Ben Ngubane, and the South African Embassy in Beijing- particularly Dumisani Rasheleng.

At the outset a small group of believers helped me with the work of our three registered charities in the UK, US and Hong Kong: Gerry Ball, Sarah Emery, Robert Hanson, Claudine Kolle, Gerry Lewis, George Liu, Becky McCane, Cory Meacham, Mike Pitcher, Ling Qiao, David K Thomas, Hugh Webster and David Witts. Some have continued their involvement.

We gradually recruited more core volunteers who dedicated their time and resources selflessly to SCT and shared the sorrows and triumphs we experienced, without any pay: Steven Prassas, Jonathan Shepherd, Steven Smith, Zhiming Xu and Dickson Yewn, etc.

Many scientists supported us along the way, many without compensation: Dr Jeremy Anderson, Dr. Richard Burroughs, Dr. Peter Crawshaw, Dr. Richard Davis, Dr Joseph van Heerden, Dr. Henry Lee, Dr. Hector Magome, Dr. Laurie Marker, Mr. Nick Marks, Dr. Jim Sanderson, Dr. David Smith, Dr. Marc Stalmans, Dr. Carl Traeholt and Dr. Nobby Yamaguchi. I am particularly thankful to Dr. Gary Koehler, who believed in my dream from the very beginning and provided steadfast advice when the charity was taking its toddler steps in the complicated world of conservation.

I placed faith in the hands-on conservation skills of South Africa to help realise my plan. The leadership of Petri Viljoen made it possible while Gus Van Dyk was instrumental in designing the first ever strategy of tiger 'rewilding', a word now accepted into the English vocabulary. Dr. Ian Player's wisdom and tenacity gave me the courage to face my own challenges.

South Africa has indeed lived up to its repution in wildlife conservation and

many of our supporters have since become my good friends: Craig Avnit, Brian Boswell, Hilton Button, Arnold Chatz, Cristina & Gianfranco Cicogna, Gerhard Damm, Eddie Van Eck, Mike Eustace, Michael Falls, Jens Friis, Nicky Fitzgerald, Adrian Gardiner, Eddie Keizan, Julius Koen, Chris Marais, Don McRobert, Ian Melass, Louis van der Merwe, Pieter van der Merwe, Mfanasibili Nkosi, Liam Paterson, Mike Sas-Rolfes, Leonard & Elree Seelig, Dr Kobus du Toit, Karen Trendler, Jaci and Jan Van Heerden, Catherine Warburton and Kim Wolhuter. I miss my dear friend the formidable Nora Kreher who, at a time when many were still in doubt, brought her immense emotional and practical support.

The Legendary kindness to animals in UK has certainly extended to the South China tigers and I am obliged to all who helped, which includes but is not limited to: Graham Aaronson, Louise Aspinall, Minzhi Bauer-Schlichtegroll, Christina Bellamy, Rona Birnie, Emma Bleasdale, Nigel Blundell, Suzi Bullough, Juanita Carbury, Stephen Cawston, Jonathan Clayton, Lawrence Cole-Morgan, Jude Cowgill, Sean Curtis-Ward, Paul Fitzgerald, Adrian & Paul Gardiner, Steven Gaydos, Richard Gnodde, Ben Goldsmith, Russell Jacobs, Nicolette Kwok, Matthew Lakin, Jonathan Laredo, David Lawley, Jim Leffman, Tony Maher, Tammy Marlar, Melissa McDaniel, Charlotte Metcalf, Sophie Mottram, Adam Murry, Ian Penman, Lucia Van der Post, John Pulford MBE, Jin Qian, Suzanne Reisman, Jo Roberts, Darshana Shilpi, Swazi Stokes, Chris Stothers, Nick Toksvig and John Wellington.

On the American continent SCT benefited from the assistance of many including Howard Buffet, Daniel Caithamer, Yvonne Chin, Sue D'Agostino, Cecilie Davidson Rudy D'Alessandro, Bill Gordon, Glenn Janss, Karen Karp, Mari Katsunuma, Dahlia Radley-Kingsley, Charlie Knowles, Vance Martin, Herold Maxwell and Nancy Sipos.

Elsewhere in the world counted supporters such as Didac Artes, Patrick Couzinet, Gilles de Dumast, John Filippakis, Laurence Kimmel, David Leffman, Andrea Manzitti, Andrew McDermott, Massimo della Ragione, Lynn Santer and James Synge, etc.

In tiger range countries countless people have donated their time and resources among them, Ivan Chan, Kirsten Conrad, Jacqui Donaldson, Cherry Yingzhi Gao, Jun Gong, Peter Gordon, James Heimowitz, Jeremy Higgs, Andrew R.R. Hirst, Steven Ho, Calvin Hui, Isa Ip, Thelma Kwan, Vincent Kwok, Anna Leung, Emily Luk, Raphael le Masne, Barun Mitra, Peco Ng, Dee Poon, James Riley, Benny Sea, John Singh, Andrew Smith, Mark Stevens, Mark Sung, Henry Winter, Kitty Wong, Susan Yeung and Kevin Zhang. A teenager by the name of Siangwei Heng from Singapore has been growing up with the charity, assisting us from the age of 13 - youngest to date!

Numerous Chinese nationals including professionals, students and government officials provided all kinds of aid. Their individual names are mentioned in the Chinese Acknowledgements.

My appreciation also goes to those artists who donated their work: Adam Binder, Steve Bindon, Hoeynn & Kei Ngu, Linda & Kevin Wain and teenager Yixia Xu. Body painting artist Craig Tracy created a stunning piece in commemoration of the Chinese Year of the Tiger and SCT's ten year anniversary. Photographer Paul Hilton's memorable images are included in this book.

The media has been particularly kind to us. Without any dedicated PR staff or agency, we rely on word of mouth. I am thankful for all who have taken an interest in our project and helped us beyond their call of duty. A list of some media that have covered our project can be found on our website. But I want to give a special mention to Reuters who followed closely our progress, and to two reporters from Xinhua News Agency - Ming Chen and Ye Yuan, whose incredible photos are also included in this book.

Besides our core team, my old friends Yanling Liu, Alan Nie, Justin Wintle and new friend Anthony Frewin helped with this book directly, while Clare Mellor was a delight to work with throughout the design process. While most photos were taken by SCT ground team, some were shot by myself and visiting friends, e.g. James Fitzgerald. All captions are based on actual events as recorded in the daily tiger monitoring diaries by my staff and myself.

Lastly, I want to thank our core team, past and present, for having confidence in me and for putting up with someone who does not take No for answer. Without their dedication, I would be a China tiger without teeth.

鸣谢

拯救中国虎国际基金会（SCT）于2000年在英国成立。若没有来自世界各行各业众多朋友与志愿者的慷慨帮助，基金会的事业只是空谈。我跟很多人其实素不相识，还把不少的朋友拉下水，"强迫"他们参与这项事业。在此我对他们不懈的支持表示由衷的感谢。

首先，我要对我丈夫斯图尔特·博锐说声"谢谢"，没有他个人对项目在资金上无条件的支持，基金会就不可能启动。之后他继续通过他建立的国际金融保护公司（CFI）为我们提供坚强的金融后盾，得以使SCT能够采取大胆创新的举动。此外，还有不少其他的企业通过赞助或无偿服务对我们提供了后援，如：绿龙基金、国泰航空公司、亚洲虎集团、香港怡东酒店、美国美邦律师事务所、天地之心珠宝、德高集团、上海滩服饰、老虎轮胎集团、丛马儿童玩具公司、触动传媒、乐趣旅游集团、海威德标本公司等等。

如珍妮·弗布恩等许多人的个人捐赠给基金会提供了坚实的后援。邓永锵爵士为我们举行的资金筹集晚宴仍然让人记忆犹新。还值得一提的是简·勒夫尔带领亨利八世预备学校的小学生们为SCT募捐的活动。

我们呼吁人的慷慨和奉献在引起公众注意方面发挥了至关重要的作用。多年来，我们获得了下列知名人士的帮助：赵忠祥、成龙、杨紫琼、陈凯歌、姜文、蒋怡、英达、郎朗、尼克·罗斯、路易斯·穆迪和克里斯多夫·艾申巴赫大师。成龙亲自驾车来参加他的虎脸发布会，让我们每个人都非常感动。

我还要感谢我弟弟全奇，因为他的鼓励我才敢贸然打电话给国家林业局。说到这，我要感谢那些有远见、有勇气，把中国虎的利益放在自己事业之上的敢做敢当的政府官员，包括：卓榕生、王维胜、王伟、刘世荣、肖文发、陆军、张希武及严旬等等。中国驻伦敦使馆的文化处，为基金会建会的发布活动提供了场所。中国驻南非前任大使刘贵今对华南虎特别关心，在他任职期间每次都长途跋涉迎接到来的小虎们。包括北开普省自然保护署、南非前文化部长本恩古巴内博士和南非驻华使馆在内的南非政府部门，也对我们予以了关怀与协助。

最初帮我在英国、美国和中国香港地区三个注册慈善机构操作管理的，是一小撮朋友，如大卫·托马斯、罗伯特·汉森、刘江、贝琦、迈凯因、乔凌、克劳灯·考勒、格里·保尔和麦克·皮彻等等。有些人直到如今还一直坚持着。

渐渐地更多的核心志愿者加入了我们的团队，他们将自己的时间和资源无私地奉献给了华南虎，与我们同甘共苦，没有任何报酬甚至还倒贴。他们中有许志明、翁狄森、林花苗、张琳、王亮、乔纳森·谢泼德、史蒂芬·史密斯和史蒂芬·布莱斯。

众多科学家支持并参与了我们的项目，往往没有任何薪俸，如：大卫·史密斯博士、卡尔·揣浩特博士、海科特·马高眉博士、吉姆·桑德森博士、彼得·科拉绍博士、劳睿·马克博士、胡德夫博士、陆厚基博士和马建章院士等等。我要特别地感谢盖瑞·克勒博士，他从基金会的建立开始就支持我的梦想，从拯救中国虎蹒跚起步时期就给我提供指导和建议，得以让我在复杂的野生动物保护领域中摸索出一条自己的道路。

我的梦想建立在南非的野生动物保护管理技术上。皮特利·维伦的领导使我的计划成为现实，而噶斯·范戴克设计的史无前例的老虎"野化"方案是项目的关键，"Rewilding"这个词现在已进入英语词汇了。

南非人在野生动物保护方面果真名不虚传，许多支持者后来都成了我的好友。英国对动物的厚道则具传奇色彩，毫无疑问也延伸到了华南虎身上。在美国，我们也得到朋友们的拥戴，如霍华德·巴菲特帮我在美国注册慈善机构。而在虎出没的亚洲，更有无数人为我们奉献了时间和资源，如名叫黄祥伟的新加坡孩子与基金会一起茁壮成长，从13岁起就成为我们中的一员，是SCT年龄最小的志愿者。其他人的名字我在英文版致谢辞中已罗列出来。

包括职员、学生和政府官员在内的许多同胞也为我们提供了各种各样的帮助。我对他们表示衷心的感谢。如高樱智、章皿星、赵方、周勉、周示凯、许帅军、邵敬涛、崔克千、余虹、陈小东、丁文蕾、董小军、段永升、傅裕、顾锦、黄禾、韩宁、胡晓、华锦州、靳硕、贾海清、孔令雯、吕克农、罗小韵、老潘、兰天明、李勇刚、李华丽、李文亮、李巍、刘巍、廖妤婷、来健、毛丹青、马莉、宋超、宋妍、秦军、司马南、唐季荣、孙佑海、王丽、吴琼、吴宗来、王海滨、王宏汇、王劲、王静、王小江、王前进、王巍、韦小可、徐俊、徐扬、杨光、姚涛、姚少华、叶钰、尹岩、袁小玉、杨树、赵广东、张弓和张翌凯等。

在此，我也要向给我们捐赠作品的那些艺术家们表示感谢，如东北画家史君先生等。为纪念中国虎年和SCT建会10年，人体彩绘艺术家克雷格·特雷西特为我们创作的"最后的中国虎"可谓轰动世界，震撼人心。

媒体对拯救中国虎国际基金会也极为支持。SCT没有任何专职的公关人员，也不聘用任何宣传机构，靠的是口口相传。我要感谢那些关注我们的媒体朋友，很多人都超出自己的责任范围来为华南虎摇旗呐喊，从我们网站的媒体报道一栏您可以看到一些报道过我们的媒体。我尤其想提到的是自始至终密切跟踪我们项目进程的路透社和新华社，两位驻南非的前任记者陈铭和袁晔的精彩图片也收集到本书中。

除了我们的核心团队外，我的老友刘燕玲和聂伟亮等也直接参与了本书的编辑工作。书中绝大部分图片来自SCT老虎谷保护区的实地工作人员，还有部分由来访的朋友和本人所拍。文字故事全部是以多年来我和团队所记载的老虎日记为依据改编。

最后，我要感谢基金会今昔的团队。谢谢他们对我的信任，并容忍我拒绝接受"不可能"这种回答。没有他们，我就是没爪没牙的中国虎。

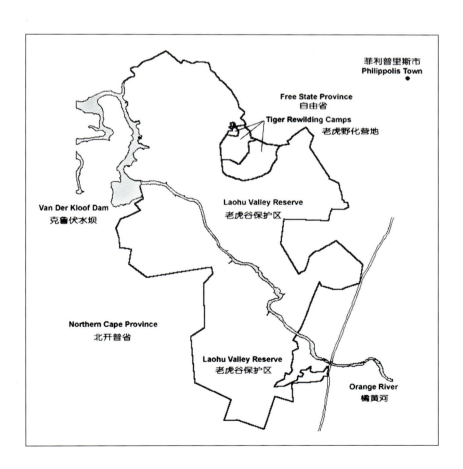

图书在版编目（CIP）数据

野化拯救华南虎：汉英对照／全莉著. — 北京：
北京出版社，2012.4
ISBN 978 - 7 - 200 - 09238 - 7

Ⅰ．①野… Ⅱ．①全… Ⅲ．①华南虎—动物保护—中
国—汉、英 Ⅳ．①Q959.838

中国版本图书馆 CIP 数据核字（2012）第 059925 号

野化拯救华南虎（汉英对照）
YEHUA ZHENGJIU HUANANHU

全 莉 著
*
北京出版集团公司
北 京 出 版 社 出版
（北京北三环中路 6 号）
邮政编码：100120
网 址：www.bph.com.cn
北京出版集团公司总发行
新 华 书 店 经 销
北京华联印刷有限公司印刷
*
889毫米×1194毫米 12 开本 21印张 387千字
2012 年 4 月第 1 版 2012 年 4 月第 1 次印刷
ISBN 978 - 7 - 200 - 09238 - 7
定价：268.00 元
质量监督电话：010 - 58572393